Water and Oil Repellent Surfaces

Water and Oil Repellent Surfaces

Editor

Ioannis Karapanagiotis

MDPI • Basel • Beijing • Wuhan • Barcelona • Belgrade • Manchester • Tokyo • Cluj • Tianjin

Editor
Ioannis Karapanagiotis
University Ecclesiastical Academy of Thessaloniki
Greece

Editorial Office
MDPI
St. Alban-Anlage 66
4052 Basel, Switzerland

This is a reprint of articles from the Special Issue published online in the open access journal *Coatings* (ISSN 2079-6412) (available at: https://www.mdpi.com/journal/coatings/special_issues/WOSurfaces).

For citation purposes, cite each article independently as indicated on the article page online and as indicated below:

LastName, A.A.; LastName, B.B.; LastName, C.C. Article Title. *Journal Name* **Year**, *Article Number*, Page Range.

ISBN 978-3-03943-541-8 (Hbk)
ISBN 978-3-03943-542-5 (PDF)

© 2020 by the authors. Articles in this book are Open Access and distributed under the Creative Commons Attribution (CC BY) license, which allows users to download, copy and build upon published articles, as long as the author and publisher are properly credited, which ensures maximum dissemination and a wider impact of our publications.

The book as a whole is distributed by MDPI under the terms and conditions of the Creative Commons license CC BY-NC-ND.

Contents

About the Editor vii

Ioannis Karapanagiotis
Water- and Oil-Repellent Surfaces
Reprinted from: *Coatings* **2020**, *10*, 920, doi:10.3390/coatings10100920 1

Ryszard Kapica, Justyna Markiewicz, Ewa Tyczkowska-Sieroń, Maciej Fronczak, Jacek Balcerzak, Jan Sielski and Jacek Tyczkowski
Artificial Superhydrophobic and Antifungal Surface on Goose Down by Cold Plasma Treatment
Reprinted from: *Coatings* **2020**, *10*, 904, doi:10.3390/coatings10090904 3

Qingwen Ma and Sihan Liu
Effect on Silt Capillary Water Absorption upon Addition of Sodium Methyl Silicate (SMS) and Microscopic Mechanism Analysis
Reprinted from: *Coatings* **2020**, *10*, 724, doi:10.3390/coatings10080724 19

Aikaterini Chatzigrigoriou, Ioannis Karapanagiotis and Ioannis Poulios
Superhydrophobic Coatings Based on Siloxane Resin and Calcium Hydroxide Nanoparticles for Marble Protection
Reprinted from: *Coatings* **2020**, *10*, 334, doi:10.3390/coatings10040334 31

Mariateresa Lettieri, Maurizio Masieri, Mariachiara Pipoli, Alessandra Morelli and Mariaenrica Frigione
Anti-Graffiti Behavior of Oleo/Hydrophobic Nano-Filled Coatings Applied on Natural Stone Materials
Reprinted from: *Coatings* **2019**, *9*, 740, doi:10.3390/coatings9110740 39

Sheng Lei, Xinzuo Fang, Fajun Wang, Mingshan Xue, Junfei Ou, Changquan Li and Wen Li
A Facile Route to Fabricate Superhydrophobic Cu_2O Surface for Efficient Oil–Water Separation
Reprinted from: *Coatings* **2019**, *9*, 659, doi:10.3390/coatings9100659 59

Doeun Kim, Arun Sasidharanpillai, Ki Hoon Yun, Younki Lee, Dong-Jin Yun, Woon Ik Park, Jiwon Bang and Seunghyup Lee
Assembly Mechanism and the Morphological Analysis of the Robust Superhydrophobic Surface
Reprinted from: *Coatings* **2019**, *9*, 472, doi:10.3390/coatings9080472 69

About the Editor

Ioannis Karapanagiotis (Professor) obtained his Ph.D. in Materials Science and Engineering from the University of Minnesota, USA, and his diploma in Chemical Engineering from the Aristotle University of Thessaloniki, Greece. He serves as an editorial board member and reviewer for several journals (more than 100), and he has published multiple research papers (more than 150) in peer-reviewed journals, books and conference proceedings. Dr. Karapanagiotis specializes in interfacial engineering and its applications in the protection and conservation of cultural heritage and in the physicochemical characterization and analysis of cultural heritage materials that are found in historic monuments, paintings, icons, textiles, and manuscripts. Dr. Karapanagiotis is a professor in the Department of Management and Conservation of Ecclesiastical Cultural Heritage Objects, University Ecclesiastical Academy of Thessaloniki, Greece.

Editorial

Water- and Oil-Repellent Surfaces

Ioannis Karapanagiotis

Department of Management and Conservation of Ecclesiastical Cultural Heritage Objects, University Ecclesiastical Academy of Thessaloniki, 54250 Thessaloniki, Greece; y.karapanagiotis@aeath.gr

Received: 21 September 2020; Accepted: 24 September 2020; Published: 25 September 2020

In the last two decades, materials of extreme wetting properties have received significant attention, as they offer new perspectives providing numerous potential applications. Water- and oil-repellent surfaces can be used, for instance, in the automobile, microelectronics, textile and biomedical industries, in the protection and preservation of constructions, buildings and cultural heritage and in several other applications relevant to self-cleaning, biocide treatments, oil–water separation and anti-corrosion, just to name a few.

The papers included in the Special Issue "Water- and Oil-Repellent Surfaces" present innovative production methods of advanced materials with extreme wetting properties which are designed to serve some of the abovementioned applications. Moreover, the papers explore the scientific principles behind these advanced materials and discuss their applications to different areas of coating technology. In particular:

Kapica et al. developed a two-step plasma modification process to create an artificial superhydrophobic surface on goose down. Two types of precursors for plasma-enhanced chemical vapor deposition (PECVD) were applied. The effects of the precursors on the wetting properties, surface morphology and chemical structure on the produced surfaces were investigated using a variety of different microscopic and spectroscopic techniques. The surface of the goose down became superhydrophobic after the plasma process and revealed a very high resistance to fungi.

Ma and Liu studied the effect of sodium methyl silicate (SMS) on the capillary water rise in silt. It was shown that SMS can effectively inhibit the rise of capillary water in silt: the maximum height of capillary rise can be reduced to 0 cm, provided that an appropriate concentration of SMS is used. SMS forms a water-repellent membrane by reacting with water and carbon dioxide, resulting in a large (120°) contact angle of water drops on treated silt. The membrane reduces the apparent surface energy of the treated silt and, moreover, it is combined with small particles of the soil, thus affecting the pores and inhibiting the rise of capillary water.

Chatzigrigoriou et al. produced calcium hydroxide nanoparticles ($Ca(OH)_2$) which were dispersed in an aqueous emulsion of silanes/siloxanes. The dispersion was deposited on marble surfaces, which obtained water repellent properties. Moreover, it was shown that the siloxane + $Ca(OH)_2$ composite coating offers good protection against water penetration by capillarity and has a small effect on the aesthetic appearance of marble. Because $Ca(OH)_2$ is chemically compatible with limestone-like rocks, which are the most common stones found in buildings and objects of tcultural heritage, the produced composite coatings have the potential to be used for conservation purposes.

Lettieri et al. produced a highly hydrophobic and oleophobic nano-filled coating using fluorine resin and silica (SiO_2) nanoparticles. The anti-graffiti performance of the coating on calcareous stones, which have been used in buildings of cultural heritage, was evaluated. For comparison, two commercial coatings were included. It was found that the protective coatings facilitated the removal of an acrylic spray paint, but high oleophobicity or paint repellence did not guarantee a complete cleaning. The stain from a felt-tip marker was difficult to remove. The cleaning with a solvent promoted the movement of the applied polymers and paint in the porous structure of the stone substrate.

Lei et al. produced a superhydrophobic copper oxide (Cu_2O) mesh through a facile chemical reaction between copper mesh and hydrogen peroxide solution without any low surface reagents treatment. The new material was designed to be used for oil–water separation. With the advantages of simple operation, short reaction time, and low cost, the produced superhydrophobic mesh showed excellent oil–water selectivity for many organic solvents. Furthermore, the Cu_2O mesh showed excellent durability, as it can be reused for oil–water separation with a high separation ability of above 95%.

Kim et al. prepared functionalized silica (SiO_2) nanoparticle dispersions which were sprayed onto acrylate-polyurethane (PU) on solid substrates. PU played the role of the binder between the thin SiO_2 layer and the substrate. The influence of the SiO_2/PU ratio on the wetting properties and the robustness of the developed surface was systematically analyzed. The best SiO_2/PU ratio to achieve durable superhydrophobicity was found to vary within 0.9 and 1.2. The evolution of the morphology of the surface with respect to the wetting properties was investigated in detail using different weight ratios of the particles to the binder. Moreover, it was concluded that the binder plays a key role in controlling the surface roughness and superhydrophobicity through the capillary mechanism.

Funding: This research received no external funding.

Conflicts of Interest: The author declares no conflict of interest.

© 2020 by the author. Licensee MDPI, Basel, Switzerland. This article is an open access article distributed under the terms and conditions of the Creative Commons Attribution (CC BY) license (http://creativecommons.org/licenses/by/4.0/).

Article

Artificial Superhydrophobic and Antifungal Surface on Goose Down by Cold Plasma Treatment

Ryszard Kapica [1], Justyna Markiewicz [2], Ewa Tyczkowska-Sieroń [3], Maciej Fronczak [1], Jacek Balcerzak [1], Jan Sielski [1] and Jacek Tyczkowski [1,*]

1. Department of Molecular Engineering, Faculty of Process and Environmental Engineering, Lodz University of Technology, Wolczanska Str. 213, 90-924 Lodz, Poland; ryszard.kapica@p.lodz.pl (R.K.); maciej.fronczak@p.lodz.pl (M.F.); jacek.balcerzak@p.lodz.pl (J.B.); jan.sielski@p.lodz.pl (J.S.)
2. Research and Innovation Centre Pro-Akademia, Innowacyjna Str. 9/11, 95-050 Konstantynów Łódzki, Poland; markiewiczjj@gmail.com
3. Department of Biology and Parasitology, Medical University of Lodz, Zeligowski Str. 7/9, 90-752 Lodz, Poland; ewa.tyczkowska-sieron@umed.lodz.pl
* Correspondence: jacek.tyczkowski@p.lodz.pl

Received: 14 August 2020; Accepted: 17 September 2020; Published: 20 September 2020

Abstract: Plasma treatment, especially cold plasma generated under low pressure, is currently the subject of many studies. An important area using this technique is the deposition of thin layers (films) on the surfaces of different types of materials, e.g., textiles, polymers, metals. In this study, the goose down was coated with a thin layer, in a two-step plasma modification process, to create an artificial superhydrophobic surface similar to that observed on lotus leaves. This layer also exhibited antifungal properties. Two types of precursors for plasma enhanced chemical vapor deposition (PECVD) were applied: hexamethyldisiloxane (HMDSO) and hexamethyldisilazane (HMDSN). The changes in the contact angle, surface morphology, chemical structure, and composition in terms of the applied precursors and modification conditions were investigated based on goniometry (CA), scanning electron microscopy (SEM), Fourier-transform infrared spectroscopy in attenuated total reflectance mode (FTIR-ATR), and X-ray photoelectron spectroscopy (XPS). The microbiological analyses were also performed using various fungal strains. The obtained results showed that the surface of the goose down became superhydrophobic after the plasma process, with contact angles as high as 161° ± 2°, and revealed a very high resistance to fungi.

Keywords: plasma deposition; organosilicon thin layers; morphology analysis; surface molecular structure; goose down; wettability; fungus resistance

1. Introduction

Plasma-enhanced chemical vapor deposition (PECVD) is a coating process widely used to produce thin layers on the surfaces of various types of materials from a broad range of precursors [1,2]. The wide application possibilities of such layers result from the fact that this method is environmentally clean, cost-effective, flexible, and the most insensitive to the shape and chemical composition of the substrate [3]. Particular attention should be paid to the layers plasma deposited from organosilicon precursors—which, due to their unique properties—have been applied in such areas as protective coatings [4,5], hydrophobic layers [6], optical coatings [7], or biocompatible films [8]. In the group of organosilicon precursors, hexamethyldisiloxane (HMDSO) and hexamethyldisilazane (HMDSN) are often preferred for the PECVD process due to their low cost, high vapor pressure, as well as obtaining stable and well-adherent layers.

There are many reports of hydrophobic or superhydrophobic layers formed on relatively flat surfaces using low-temperature non-equilibrium (cold) plasma and organosilicon precursors [9–14].

However, this type of plasma has only occasionally been reported for use in modifying natural, fluffy materials such as goose down [15,16]. Since the goose down is considered as the best filling material in textile products [17], due to the excellent thermal insulation and fill power (loft) [18–20], research has begun towards the elimination of the main drawbacks of down, i.e., moisture absorption and mold growth. These drawbacks lead to an almost complete loss of the thermal insulation properties of down with a loss of loft and a significant deterioration of its suitability, as well as environmental health risk by opening the way for the growth and development of allergenic (harmful) fungi [21–23]. This is due to the removal of waxes and oils from down in the purification process, which secures it naturally [24].

It seems that the best solution, in this case, would be to deposit a thin, highly hydrophobic layer on the down feathers, which would protect them against wettability. Presumably, such a coating could also be antifungal and antibacterial [25,26]. Recent reports show that the surface hydrophobicity for a given coating is crucial for these properties [27,28].

In this study, we have attempted to produce an effective superhydrophobic and antifungal coating on the surface of goose down feathers by the PECVD method using organosilicon precursors, such as HMDSO and HMDSN.

2. Materials and Methods

2.1. Materials

The base material for the research was the white goose down, designated as type: 90/10 (approx. 90% of plumes and the rest, approx. 10%, of down feathers), 750 CUIN (down elasticity, indicating the volume in cubic inches that one ounce of down occupies), and 1000 mm (translucency, i.e., the height of the water column in which an ounce of fluff was shaken, through which the bottom of the measuring vessel can be seen), provided by Animex Foods Comp. (Dobczyce, Poland). For investigations, the down was used in the native form or as pellets produced by its grinding in a mill (Vertical Laboratory Planetary Ball Mill model XQM-16A: 2.0 L jar, alumina balls with a diameter of 5–15 mm and a total weight of 0.5 kg, down load of 3 g per run; TENCAN, Changsha, China) followed by pressing in a hydraulic press (Atlas 15T model, SPECAC, Orpington, UK) using 0.17 g of ground down for each pellet and a pressure of 15 bar for 5 min at room temperature. In this way, down pellets with a diameter of 13.0 mm, a thickness of 2.0 mm, and a weight of 0.15 g were formed. Due to the much greater uniformity of the surface and therefore much easier measurement of the contact and tilt angles, the pellets were used as a reference substrate to optimize plasma processes in terms of obtaining the best hydrophobic properties of the down material surface. Figure 1 shows different forms of the tested down.

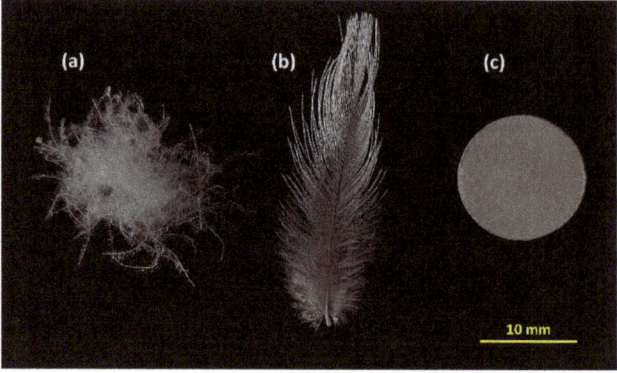

Figure 1. Forms of the tested down: (**a**) plume; (**b**) down feather; (**c**) pellet.

For study of the molecular structure of the plasma deposited layers and its influence on the hydrophobicity of the surface of these layers, while minimizing the contribution of the surface morphology in this effect, ultra-smooth silicon wafers (Institute of Electronic Materials Technology, Warsaw, Poland) were used as a substrate. The silicon wafers with thickness >2 mm and orientation <100> were characterized by a tolerance of thickness, flatness, and total thickness variation (TTV) <5 µm and surface roughness of <0.5 nm.

2.2. Thin Layer Deposition

A laboratory capacitively coupled RF glow discharge (13.56 MHz) plasma reactor was used to perform both surface activation of the samples and the deposition of thin layers. Two electrodes with an area of 64 cm^2 each and a 3 cm gap between them were placed in parallel in the reactor chamber connected to a pumping system. The bottom powered electrode also supported the substrates in the form of pellets and silicon wafers. In the case of native down, to ensure complete and uniform coverage of all down elements, the material was stirred with a suitable agitator inside the plasma reactor. Reaction gases (in flow) were introduced through the upper perforated grounded electrode. First, the samples were activated with argon plasma generated in Ar (99.999% purity, Linde Gas, Cracow, Poland) with a flow of 4 sccm, which was stabilized by a mass flow controller SLA 5850 (Brooks Instrument BV, Veenendaal, The Netherlands). The initial pressure in the reactor chamber was approx. 8 Pa. Then, the deposition process was carried out with HMDSO (≥98% purity, Merck KGaA, Darmstadt, Germany) or HMDSN (≥98% purity, Merck KGaA, Darmstadt, Germany) as precursors. Their flow rates (approx. 0.35 sccm at the initial pressure in the chamber of 3.2 Pa) were controlled by thermal stabilization of precursor containers at 0 °C and precision leak valves. The applied discharge power was 25–80 W. The Ar plasma treatment time was 30–60 s, while the deposition time was 60–240 s. The selected parameters ensured stable plasma discharge conditions throughout the process. The layer thicknesses were determined for the layers deposited on the silicon wafers by the interference method using an interference microscope Nikon Eclipse LV 150N (Nikon, Tokyo, Japan).

Hereinafter, the plasma-deposited layers from HMDSO and HMDSN will be referred to as pp-HMDSO and pp-HMDSN layers, respectively.

2.3. Hydrophobicity Measurements

The hydrophobicity was determined by measuring the water contact angle (CA) at room temperature (25 °C) and using deionized water (Millipore Direct-Q 3 UV system, Millipore SAS, Molsheim, France). The analysis was carried out by an optical goniometer Theta 2000 (KSV Instruments Ltd., Helsinki, Finland) equipped with an automated table and liquid dispenser, ensuring high repeatability of measurements. A drop of water (4 µL) was placed on the surface of a given sample and the static contact angle was measured. The procedure was repeated for a minimum of 10 samples prepared under the same conditions to determine the average CA value. Figure 2 shows an example of photographs of water droplets placed on the surface of the down pellet and plume after coating by plasma-deposited HMDSN, with determined contact angles. The tilt angle was also determined by running at least 10 trials for a given sample using a 4 µL water drop.

For the visual analysis of the down hydrophobicity before and after the plasma deposition process, the water shake test was used [29]. The behavior of the down was observed after its vigorous shaking with water for 10 s.

2.4. Molecular Structure Analysis

Two spectroscopic techniques were employed to study the chemical structure of the plasma deposited layers, namely Fourier transform infrared-attenuated total reflectance spectroscopy (FTIR-ATR) and X-ray photoelectron spectroscopy (XPS).

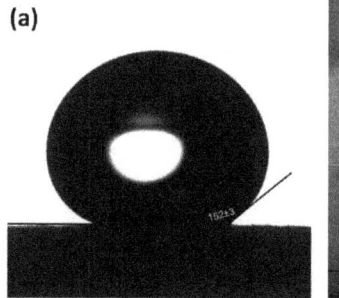 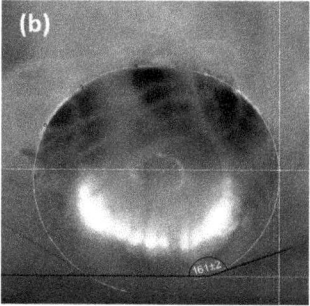

Figure 2. Photographs of water droplets on surfaces covered with pp-HMDSN, with designated contact angles: (**a**) down pellet; (**b**) plume.

A Jasco FTIR 4200 spectrometer equipped with an ATR sampling accessory (Multi-Reflection ATR PRO410-M, angle of incidence 45° and ZnSe prism) (Jasco Inter. Co., Ltd., Tokyo, Japan) was used for FTIR-ATR studies that were performed on the samples prepared on the silicon wafers. To eliminate disturbances caused by water vapor and carbon dioxide from the air, the measuring system was purged with dry nitrogen (99.999% purity; Linde Gas, Cracow, Poland). Spectra were recorded in the range of 4000–700 cm^{-1} with a resolution of 4 cm^{-1}. The scanning rate was set at 1 cm/s. As standard, 50 scans for each spectrum were collected and averaged. As-recorded spectra were calibrated by subtraction of the substrate response and taking into account the thickness of the layers.

For XPS investigations, an AXIS Ultra DLD (Kratos Analytical Ltd., Manchester, UK) spectrometer was utilized, equipped with a monochromatic Al-Kα X-ray source (150 W, 1486.6 eV) with the spot size 300 × 700 µm. The base pressure in the analytical chamber was approx. 5×10^{-8} Pa; the pass energy and step size were set to 20 and 0.1 eV, respectively, for all high-resolution measurements. The XPS spectra were calibrated against the C 1s peak assigned to the C–C/C–H bonds and positioned at 284.8 eV. Due to the insulating nature of the deposited layers, which were prepared both on the native down and silicon wafers, charge neutralization was used when recording the XPS spectra.

2.5. Surface Morphology Studies

The surface morphology of the down pellets, plumes, and silicon wafers before and after the plasma deposition process was investigated by scanning electron microscopy (SEM) using an FEI Quanta 200F microscope (Thermo Fisher Scientific, Hillsboro, OR, USA), equipped with a field emission gun (FEG). Due to the insulating nature of samples, all SEM analyses were performed under a nitrogen atmosphere at a pressure of 100 Pa (low vacuum operating mode). This mode avoids coating the sample with a thin conductive layer, such as carbon or gold, which is important in order not to distort the surface topography. ImageJ software (U.S. National Institutes of Health, Bethesda, MD, USA) was used to determine the size of globular structures present on the surface of down pellets and plumes with deposited plasma layers.

2.6. Microbiological Testing

The microbiological tests consisted of two stages. In the first stage, the down obtained from worn and dirty down products (down jackets, sleeping bags) was investigated. The down was cultured on Sabouraud liquid and solid media with chloramphenicol and gentamicin (both from bioMérieux Polska, Warsaw, Poland) for isolation and cultivation of fungi. The tested material was incubated for 24 h at 35 °C, and then 7–14 days at 25 °C. In turn, single colonies were picked from the axenic strains of the isolated fungi and cultured on the Czapek-Dox medium (Sigma-Aldrich, Poznan, Poland) for the identification. The strains of isolated fungi were determined based on the macroscopic image of the grown colonies, as well as direct and colored microscopic preparations.

In the second stage, the possibility of fungal infection in fresh and clean goose down was tested. The down samples, both without plasma treatment and with the deposited layer from HMDSO or HMDSN, were placed on Petri dishes moistened with phosphate buffer solution (PBS) and the cells of the selected fungi—such as *Aspergillus fumigatus*, *Aspergillus flavus*, and *Aspergillus niger*—were introduced there in the form of a suspension in PBS solution with a concentration of 5×10^7 cells/mL (densitometric evaluated). The infected down was incubated for 24 h at 35 °C, and then 7 days at 25 °C, maintaining a constant high humidity. The degree of infection was assessed by macroscopic observation scores.

3. Results and Discussion

3.1. Superhydrophobicity

To determine the most favorable parameters for the deposition of the layers in terms of their hydrophobicity, tests of the contact angle were carried out for the layers deposited on down pellets, as a reference substrate, under various conditions of the plasma process. Two types of precursors were used, i.e., HMDSO and HMDSN, and parameters such as discharge power and treatment time were changed for both the activation and deposition processes. The representative results are presented in Table 1. The most optimal parameters (red dots), regardless of the type of precursor, turned out to be the discharge power of 40 W and the treatment time of 30 s in the activation process and 25 W and 240 s in the deposition process, respectively, although the contact angle for the plasma deposit of HMDSN was slightly greater than that of HMDSO. The thickness of the layers produced from both HMDSO and HMDSN was 400–500 nm. The down pellets, which were plasma modified with these parameters, were visually indistinguishable from the unmodified ones. It should be noted, however, that the discharge power of 80 W in the activation process changes the color of the pellet surface. Also, a deposition time of 60 s is too short because water droplets on the pellet surface during CA measurements start to soak after about 90 s, even at initially high contact angles.

Table 1. Optimization of plasma process parameters in search of the best surface hydrophobicity on down pellets.

Sample No.	Plasma Activation		Plasma Deposition		Contact Angle
	Power	Time	Power	Time	
	(W)	(s)	(W)	(s)	(deg)
pp-HMDSO					
1	80	30	40	120	138 ± 1
2	80	60	40	120	133 ± 2
3	40	30	40	120	142 ± 2
4	40	60	40	120	139 ± 3
5	25	30	40	120	136 ± 4
6	25	60	40	120	138 ± 3
7	40	30	40	60	137 ± 5
8	40	30	40	240	143 ± 2
9	40	30	25	60	141 ± 4
10	40	30	25	120	143 ± 3
11	40	30	25	240	144 ± 5
pp-HMDSN					
12	40	30	25	60	143 ± 5
13	40	30	25	120	147 ± 1
14	40	30	25	240	152 ± 3

Taking the determined optimal parameters of the plasma process, the layers were deposited on the native down. The result of the water shake test is shown in Figure 3. It is evident that the plasma treatment made the down surface very hydrophobic. The untreated down is completely wetted, while the plasma coated down repels water effectively. The water shake test repeated many times does not completely change the properties of the plasma-treated down. The measured contact and tilt angles for this case and those for the down pellets treated with the same parameters of the plasma process (from Table 1) are compared in Table 2. As can be seen, the native down achieved even better hydrophobicity than down pellets. Besides the fact that the contact angles are higher, the tilt angles are close to zero. Water droplets placed on the surface of the down samples with plasma deposited layers roll down with the smallest inclination of the sample plane. The obtained results classify such a plasma-modified down as superhydrophobic. According to the currently accepted definition, superhydrophobic surfaces are those for which the water contact angle is ≥150°, and the tilt angle is ≤10° [30–32].

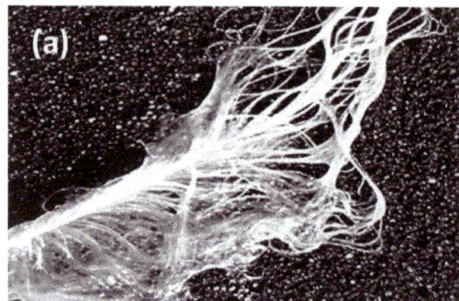 

Figure 3. Photographs of down feathers after the water shake test: (**a**) without plasm treatment; (**b**) with thin layer of pp-HMDSN.

Table 2. Contact angles, tilt angles, and the average size of globular structures on the deposited layers surface for the plasma process optimal parameters

Sample Name	Contact Angle (deg)	Tilt Angle (deg)	Globules (nm)
pellet pp-HMDSO No. 11	144 ± 5	9 ± 1	400 ± 100
pellet pp-HMDSN No. 14	152 ± 3	6.5 ± 0.5	190 ± 70
down pp-HMDSO	150 ± 2	≈0	70 ± 20
down pp-HMDSN	161 ± 2	≈0	50 ± 20
Si wafer pp-HMDSO	100 ± 2	20 ± 1	0
Si wafer pp-HMDSN	95 ± 4	25 ± 1	0

Table 2 also includes the results for the layers deposited on silicon wafers, which in turn exhibit a lower hydrophobicity than the native down and down pellets. The CA values measured for these samples showed good agreement with the literature reports for such materials obtained under similar conditions [33–36].

3.2. Surface Morphology

Based on Table 2, it can be concluded that despite the deposition of layers from the same material (the same precursor and the same parameters of the plasma process), differences in the hydrophobicity appear, which could be associated with a different surface morphology of the formed plasma layers. Indeed, as shown in Figure 4, the SEM investigations revealed a globular morphology in the case of the down pellets and the native down samples. On the other hand, the deposited layers on the silicon wafers are completely smooth at the nanoscale. The determined average size and its standard deviation for the globular structures occurring on the surface of the tested samples are given in Table 2.

Generally, it can be stated that reducing the size of globular structures increases the hydrophobicity of the surface (taking into account both the contact and tilt angle). On the other hand, a significant reduction in contact angles and an increase in tilt angles, when there is a complete lack of surface globular morphology, as in the case of layers deposited on silicon wafers, indicates a very important role of the globular structure in creating superhydrophobicity of the native down surface. This observation is in line with an already well-established understanding of superhydrophobicity. Since the research on the lotus effect, it is known that two factors—namely micro- and nanoscale hierarchical surface morphology such as those found on lotus leaves, as well as the molecular structure of the surface—are critically important for this effect [37,38].

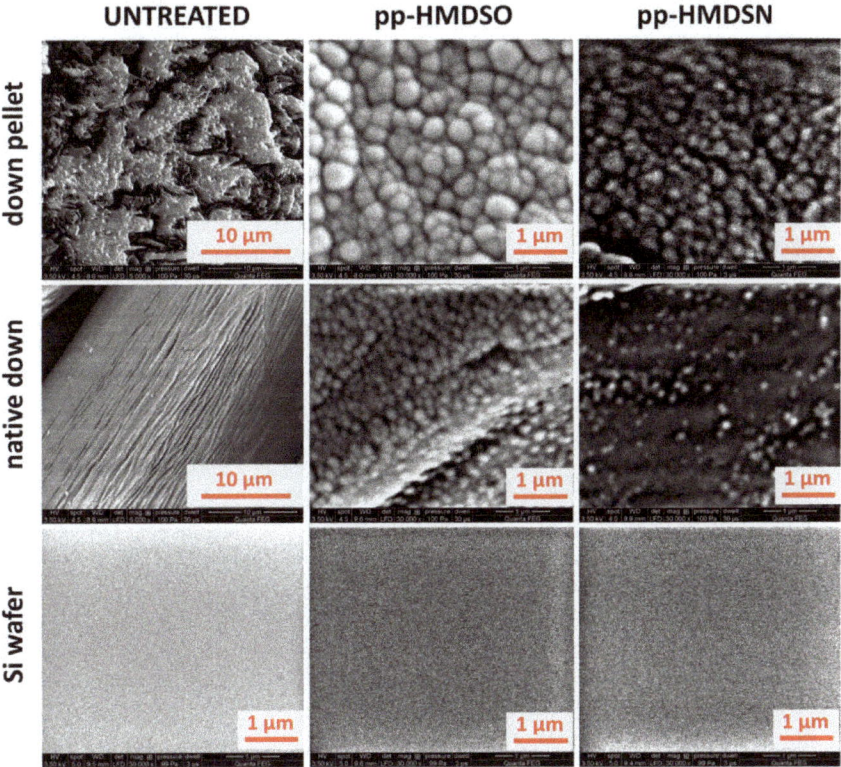

Figure 4. SEM images of the surface of various samples before and after the plasma deposition of thin layers from HMDSO and HMDSN.

3.3. Molecular Structure

The comparison of the contact and tilt angles for the pp-HMDSO and pp-HMDSN layers deposited on silicon wafers—i.e., layers without a globular structure (Table 2)—shows that their molecular structure, as already mentioned above, also influences the hydrophobic properties of the surface. To determine the basic molecular structure of the layers, mainly by identifying the functional groups that are present there, studies by FTIR-ATR spectroscopy were performed. Figure 5 shows the FTIR-ATR spectra in the representative ranges of 700–1450 and 2750–4000 cm^{-1} for the plasma-deposited layers from HMDSO and HMDSN precursors. The assignment of individual IR absorption bands to the respective chemical bonds and vibration modes, referring to the cited literature, is given in Table 3.

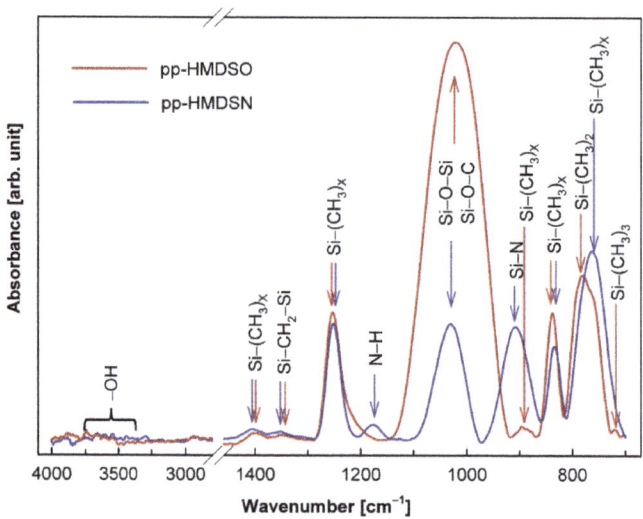

Figure 5. FTIR-ATR spectra for the pp-HMDSO and pp-HMDSN deposited on silicon wafers.

Table 3. Assignments of bands detected in FTIR-ATR spectra of pp-HMDSO and pp-HMDSN thin layers.

Layer	Band (cm^{-1})	Vibrational Mode	References
pp-HMDSO	1399	CH$_3$ asymmetric bending in Si–(CH$_3$)$_x$, (x = 1, 2, 3)	[34,38,39]
	1350	CH$_2$ scissor vibration in Si–CH$_2$–Si	[40]
	1253	CH$_3$ symmetric bending in Si–(CH$_3$)$_x$, (x = 1, 2, 3)	[33,34,39–42]
	1022	Si–O–Si asymmetric stretching; Si–O–C stretching	[33,34,39–42]
	896	CH$_3$ rocking in Si–(CH$_3$)$_2$	[40]
	834	CH$_3$ rocking in Si–(CH$_3$)$_3$	[33,34,39–41]
	781	CH$_3$ rocking in Si–(CH$_3$)$_2$, Si–O–Si bending	[33,34,39–42]
	720	CH$_3$ rocking in Si–(CH$_3$)$_3$	[40,41]
pp-HMDSN	1405	CH$_3$ asymmetric bending in Si–(CH$_3$)$_x$, (x = 1, 2, 3)	[39,43,44]
	1352	CH$_2$ asymmetric bending in Si–CH$_2$–Si	[43]
	1251	CH$_3$ symmetric bending in Si–(CH$_3$)$_x$, (x = 1, 2, 3)	[39,43–46]
	1176	N–H bending	[35,44]
	1031	Si–O–Si asymmetric stretching; Si–O–C stretching	[43,44,46]
	908	Si–N asymmetric stretching in Si–NH–Si	[39,43–46]
	835	CH$_3$ rocking in Si–(CH$_3$)$_3$	[39,43–46]
	763	Si-C stretching in Si–(CH$_3$)$_x$, (x = 1, 2, 3)	[43–45]

The spectra show a clear similarity in the molecular structure of both layers, especially in the area of Si–C bonds. Small shifts in the position of the bands for the same functional groups in these layers probably result from the difference in the chemical structure of the precursors and, consequently, some differences in, for example, the layer density, molecular environment of these functional groups, and cross-linking structure [35,42,47]. It is worth noting that despite the lack of oxygen atoms in the HMDSN precursor molecule, the IR spectrum of the layer contains a band assigned to the Si-O-Si and Si-O-C groups. This is most likely due to the oxidation reactions that take place after removing the layer from the plasma reactor chamber and contacting with air [39,43,44,46].

What is particularly important, however, is the lack of hydroxyl groups in both cases, which should be visible in the range of 3400–3650 cm^{-1}. The absence of these polar groups undoubtedly improves the hydrophobicity of the material surface. In turn, the presence of polar N-H groups in the pp-HMDSN layers, as indicated by the weak band at 1176 cm^{-1}, should act towards hydrophilicity. According to the well-established view [34,36,48], the organic fraction in the form of methyl groups derived from precursor molecules, which are retained in the deposited layers, is essential for the hydrophobic properties of these layers. As shown in Figure 5, the content of methyl groups in both materials is similar. Thus, it is not surprising that the contact angles are also close to each other (Table 2), and the slight difference can be blamed on, for example, the presence of N-H groups in the pp-HMDSN layers. However, much higher values of the contact angle and a greater difference between them were obtained when the native down was used as the substrate (Table 2). This indicates that, in this case, the surface morphology of the layers plays a critical role in the hydrophobicity and makes it possible to achieve superhydrophobic properties.

The investigations carried out by FTIR-ATR spectroscopy provided information on the average molecular structure of the entire layer because this technique analyzes the sample with a penetration depth of about 2 μm [49]. However, for hydrophobic properties, only the real molecular structure of the surface itself is important, therefore investigations were carried out using XPS spectroscopy, for which the penetration depth does not exceed 10 nm. As samples, the pp-HMDSO and pp-HMDSN layers deposited on silicon wafers were used. Typical XPS survey (wide) scans for the samples are shown in Figure 6.

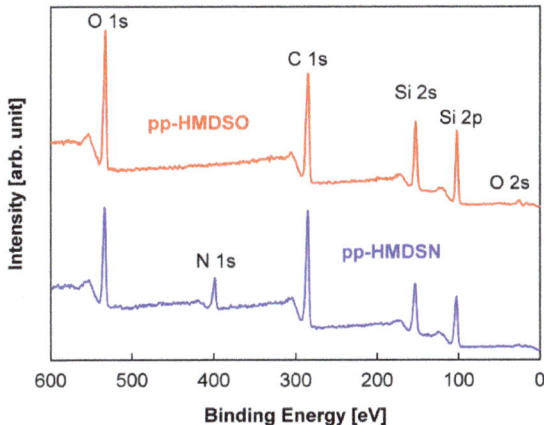

Figure 6. XPS wide scan spectra for the pp-HMDSO and pp-HMDSN deposited on silicon wafers.

The main peaks of the spectra in Figure 6 are assigned, based on the binding energy, to the core levels of the O, N, C, and Si elements. On this basis, the elemental atomic composition was determined, which is presented in Table 4. What is particularly important is the fact that the XPS analysis confirmed the presence of oxygen in the pp-HMDSN layers; however, the oxygen content in this case is much higher than the FTIR investigations showed. From the XPS (Table 4), the ratio between the oxygen content in the layers from HMDSO and HMDSN is about 1.6 (0.34/0.21), whereas from the FTIR (taking the band assigned to the Si-O-Si and Si-O-C groups in Figure 5), it is higher than 3. This indicates that oxygen is present mainly on the surface of the layers from HMDSN, which confirms its origin from oxidation (aging) processes after the layers are removed from the plasma reactor chamber.

Table 4. Elemental atomic composition of the pp-HMDSO and pp-HMDSN determined based on XPS analysis

Sample Number	Surface Composition (at %)				O/(Si + C)
	Si	C	N	O	
Si wafer pp-HMDSO	22.20	52.19	0	25.61	0.34
Si wafer pp-HMDSN	21.53	54.07	8.45	15.95	0.21

To accurately identify the chemical bonds present in the deposited layers, narrow scans of the O 1s, N 1s, C 1s, and Si 2p regions were analyzed. These core level spectra were fitted using a Gaussian convolution with a Shirley-type background. The revealed bands were assigned to the appropriate species based on the literature data concerning the XPS analysis of amorphous hydrogenated organosilicon layers deposited by the PECVD method [39,44,50–52]. Figures 7 and 8 present the developed core level spectra for the pp-HMDSO and pp-HMDSN layers, respectively.

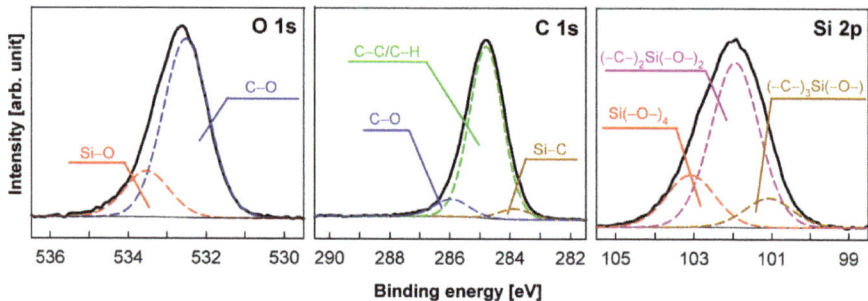

Figure 7. XPS core-level spectra of **O 1s**, **C 1s**, and **Si 2p** for the pp-HMDSO layer.

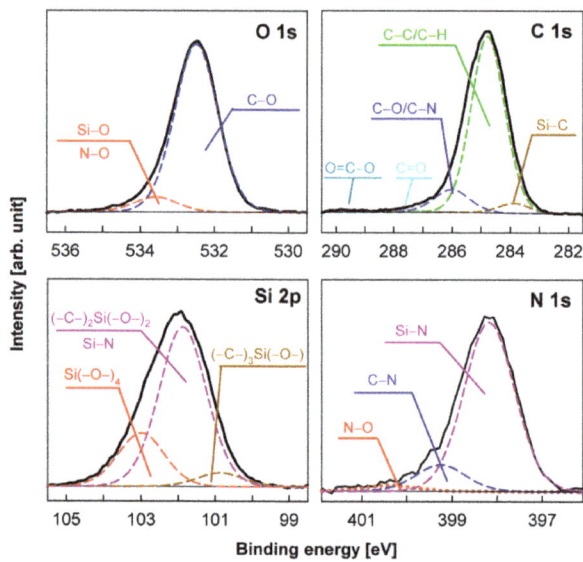

Figure 8. XPS core-level spectra of **O 1s**, **C 1s**, **Si 2p**, and **N 1s** for the pp-HMDSN layer.

The O 1s spectra were deconvoluted into two bands, regardless of the type of precursor used. The first band at 532.5 eV is attributed to the C–O bonds, and the second band at 533.6 eV is assigned to Si–O bonds and a small amount of N–O bonds for the pp-HMDSN layers. These results once again fully confirm the significant presence of oxygen bonds on the surface of the layers deposited from HMDSN. In turn, the C 1s spectra were fitted with three components for the pp-HMDSO layer and five components for the pp-HMDSN layer. Bands at 283.9 and 284.8 eV are assigned to the C–Si and C–C/C–H groups, respectively, for both types of layers. The next band at 286.0 eV occurring for pp-HMDSO corresponds to C–O bonds, and in the case of pp-HMDSN, it can be assigned to both

C–O and predicted C–N bonds. The last two very weak bands observed only for pp-HMDSN, at 287.7 and 289.6 eV, can be attributed sequentially to the C=O and O=C–O moieties, which are the result of the aforementioned oxidation (aging) processes taking place on the surface of the pp-HMDSN layers [39,44].

In the case of the Si 2p spectra, three-component bands were applied to fit it regardless of the precursor used. For the pp-HMDSN layer (Figure 8), the main band at 101.9 eV is assigned to the Si–N groups [44]. However, at the same binding energy, the (-C-)$_2$Si(-O-)$_2$ configuration can also be present [50,52]. The second band at 103.1 eV is associated with Si atoms bonded to four adjacent oxygen atoms (Si(-O-)$_4$), and the last band at 100.9 eV comes from silicon bonded to three carbon atoms and one oxygen atom, denoted as (-C-)$_3$Si(-O-). In the case of the pp-HMDSO layer (Figure 7), all three bands correspond to silicon oxycarbide species [53]. The band centered at 101.1 eV is related to (-C-)$_3$Si(-O-). The next band at 102.1 eV is assigned to (-C-)$_2$Si(-O-)$_2$, and the last band (103.1 eV) is attributed to the Si(-O-)$_4$ chemical structure, which is originated from silica in the form of SiO_2. Based on the Si 2p spectra, it can be concluded that for both types of layers, cross-linking structures in the form of Si-O and Si-C were formed, which play an important role in the superhydrophobic nature of the layer surfaces. It should be added that, in the case of pp-HMDSN, some of the cross-linking structures are also created with the participation of nitrogen atoms. The N 1s spectrum (Figure 8), fitted by three bands, complements the results presented above. The most intense band at 398.1 eV is attributed to the Si–N bond and confirms that nitrogen is mainly bonded to the Si-containing cross-linked structure. The second band associated with the C–N groups, located at 399.3 eV, shows the bond not present in the precursor molecule, indicating a high degree of fragmentation despite the low power of plasma discharge [44]. The weakest band assigned to N-O bonds (at 400.4 eV) should naturally be associated with the oxidation (aging) processes already discussed above, which occur after the sample comes into contact with air.

The XPS investigations were also carried out on samples of pp-HMDSO and pp-HMDSN layers prepared on the native down with the optimal plasma process parameters. The surface molecular structure of the samples produced in this way was practically the same as for the layers deposited on silicon wafers. An interesting conclusion, however, can be drawn from these studies when we compare the XPS wide scan spectra for plasma untreated down and down covered with the pp-HMDSO layer (Figure 9). A small amount of silicon on the uncoated surface of the down is most likely related to the contamination remaining after cleaning in the manufacturing process. What is important, however, is that the N 1s band associated with the chemical structure of down is completely absent in the sample with the pp-HMDSO layer. This indicates a homogeneous and uniform coverage of the down surface by the plasma-deposited layer.

3.4. Antifungal Properties

To determine the usefulness of down covered with the plasma-produced layer, apart from its superhydrophobicity, the influence of this layer on the antifungal properties was also examined. For this purpose, first, the down obtained from worn and dirty down products (down jackets, sleeping bags) was tested to assess the scale of the fungal problem. It was found, like Woodcock et al. [23], who had studied fungal contamination of synthetic and feather pillows, that the main strains isolated were *Aspergillus fumigatus*, *Aspergillus flavus*, and *Aspergillus niger*. Besides, we also isolated *Alternaria chlamydospora*, *Alternaria tenuissima*, as well as *Penicillium chryzogenum*, *Penicillium brevicompactum*, and *Chaetomium* spp.

Then, the susceptibility to fungal infection of fresh and clean native down, which was the subject of this paper, was investigated. The three most common strains isolated by us were selected as model fungi, namely *A. fumigatus*, *A. flavus*, and *A. niger*. Down samples, both without plasma coating and with the deposited layer from HMDSO or HMDSN, were contaminated with these fungi. Figure 10 shows, as an example, the growth of the selected fungi on samples of native down and covered with the pp-HMDSO layer. The same results were obtained for the pp-HMDSN coating. As can be seen, the uncovered down is susceptible to fungal attacks, while the down with plasma coating shows considerable resistance to fungi. More detailed studies have shown that damage spots on down feathers are particularly susceptible to fungal growth. The finely cut feathers are much more likely to be attacked by the tested fungi than the undamaged ones. Thus, the deposition of the plasma layer protects them against such infection. The second important factor is the superhydrophobicity obtained as a result of plasma layer deposition. Water repulsion and maintaining the feathers in a non-wetted state also effectively prevents the development of fungi.

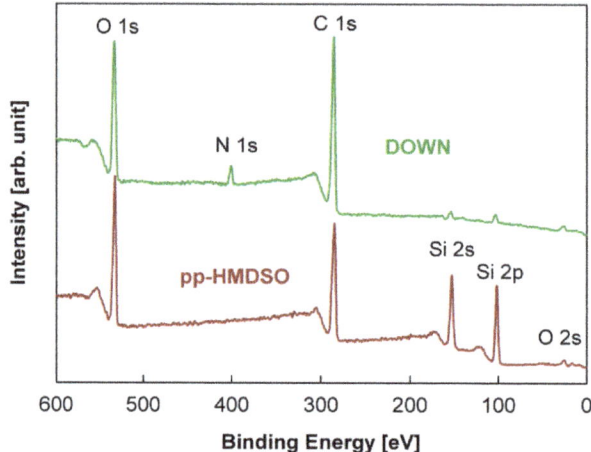

Figure 9. XPS wide scan spectra for the native down and the pp-HMDSO layer deposited on this down.

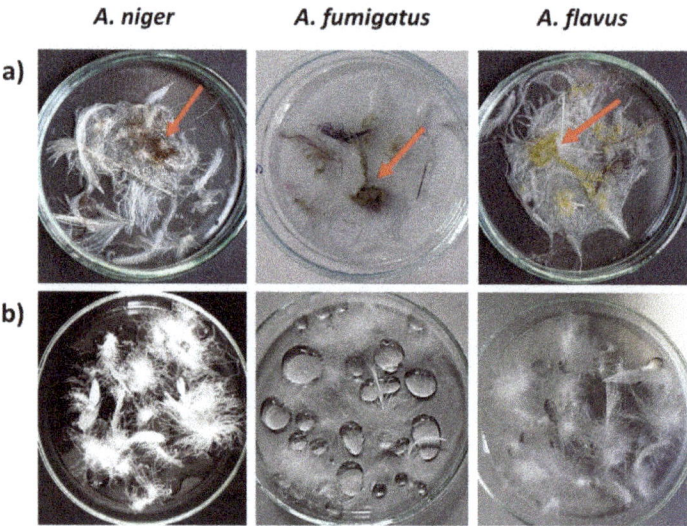

Figure 10. Microbiological examination of goose down with selected strains of fungi: (**a**) native down without plasma treatment (red arrows indicate places of fungal growth); (**b**) down coated with pp-HMDSO (no fungal growth).

4. Conclusions

The results obtained in this work have proved that cold plasma is an excellent tool in modern technology. The use of this method in a two-step process consisting of plasma activation of a surface, followed by depositing a thin layer from precursors such as HMDSO or HMDSN, has allowed creating artificial superhydrophobicity on the surface of goose down. Very high contact angles (up to 161°) and very low tilt angles (close to 0°) have been achieved. This means that the down becomes completely waterproof and repels water effectively. The superhydrophobic surface also reveals very high resistance to fungi. The deposited layers are homogeneous and uniformly cover the entire surface of the down, at the same time showing high stability, which is a consequence of the chemical bonding of the layer to the down surface as a result of the plasma process. It should also be added that the measurements of the wettability repeated several times after longer periods of time (months) did not show any changes in the hydrophobic properties of the plasma treated down.

The studies conducted on the plasma-deposited layers from HMDSO and HMDSN confirm that both the molecular structure and the surface morphology have a significant influence on superhydrophobicity. Both of these factors can be controlled in a wide range by choice of precursors and parameters of plasma processes, which strengthens the view of high usefulness of the PECVD method and should ensure its permanent strong position among the methods used to create superhydrophobic surfaces [54].

The superhydrophobic down produced in the two-stage plasma process has great potential for practical application. This is confirmed by successful tests, for example, conducted by the K2 – Polish National Winter Expedition in 2017/18, which used down equipment made of our superhydrophobic down.

5. Patents

Some parts of the technology for producing a superhydrophobic and antifungal artificial surface on goose down presented in this paper have already been patented [16].

Author Contributions: R.K. writing the manuscript and contribution to the thin layers deposition; J.M. contribution to the hydrophobicity measurements; E.T.-S. contribution to the microbiological tests; M.F. contribution to the FTIR-ATR analysis; J.B. contribution to the XPS studies; J.S. contribution to the surface morphology studies; J.T. conceptualization, supervision, writing and editing the manuscript. All authors have read and agreed to the published version of the manuscript.

Funding: This work was financially supported by the Polish National Centre for Research and Development (NCBiR), project number PBS1/B9/16/2012, and by the Ministry of Science and Higher Education (MNiSW) based on an agreement number MNiSW/2017/DIR/32.

Acknowledgments: The authors would like to thank K. Skalska and K. Wrześniewska-Tosik for their technical assistance.

Conflicts of Interest: The authors declare no conflict of interest.

References

1. Wrobel, A.M.; Wertheimer, M.R. Plasma-polymerized organosilicones and organometallics. In *Plasma Deposition, Treatment, and Etching of Polymers*; d'Agostino, R., Ed.; Academic Press: New York, NY, USA, 1990; pp. 163–268.
2. Biederman, H. (Ed.) *Plasma Polymer Films*; Imperial College Press: London, UK, 2004.
3. Liu, K.; Tian, Y.; Jiang, L. Bio-inspired superoleophobic and smart materials: Design, fabrication, and application. *Prog. Mater. Sci.* **2013**, *58*, 503–564. [CrossRef]
4. Zajickova, L.; Bursikova, V.; Kucerova, Z.; Franta, D.; Dvorak, P.; Smid, R.; Perina, V.; Mackova, A. Deposition of protective coatings in rf organosilicon discharges. *Plasma Sources Sci. Technol.* **2007**, *16*, S123–S132. [CrossRef]

5. Fracassi, F.; d'Agostino, R.; Palumbo, F.; Angelini, E.; Grassini, S.; Rosalbino, F. Application of plasma deposited organosilicon thin films for the corrosion protection of metals. *Surf. Coat. Technol.* **2003**, *174–175*, 107–111. [CrossRef]
6. Behnisch, J.; Tyczkowski, J.; Pela, I.; Hollander, A.; Ledzion, R. Formation of hydrophobic layers on biologically degradable polymeric foils by plasma polymerization. *Surf. Coat. Technol.* **1998**, *98*, 872–874. [CrossRef]
7. Martinu, L.; Poitras, D. Plasma deposition of optical films and coatings: A review. *J. Vac. Sci. Technol. A* **2000**, *18*, 2619–2645. [CrossRef]
8. Favia, P.; d'Agostino, R. Plasma treatments and plasma deposition of polymers for biomedical applications. *Surf. Coat. Technol.* **1998**, *98*, 1102–1106. [CrossRef]
9. Hirotsu, T.; Tagaki, C.; Partridge, A. Plasma copolymerization of acrylic acid with hexamethyldisilazane. *Plasmas Polym.* **2002**, *7*, 353–366. [CrossRef]
10. Azioune, A.; Marcozzi, M.; Revello, V.; Pireaux, J.-J. Deposition of polysiloxane-like nanofilms onto an aluminium alloy by plasma polymerized hexamethyldisiloxane: Characterization by XPS and contact angle measurements. *Surf. Interface Anal.* **2007**, *39*, 615–623. [CrossRef]
11. Twardowski, A.; Makowski, P.; Małachowski, A.; Hrynyk, R.; Pietrowski, P.; Tyczkowski, J. Plasma treatment of thermoactive membrane textiles for superhydrophobicity. *Mater. Sci. Medzg.* **2012**, *18*, 163–166. [CrossRef]
12. Mahlberg, R.; Niemi, H.E.-M.; Denes, F.; Rowell, R.M. Effect of oxygen and hexamethyldisiloxane plasma on morphology, wettability and adhesion properties of polypropylene and lignocellulosic. *Int. J. Adhes. Adhes.* **1998**, *18*, 283–297. [CrossRef]
13. Mobarakeh, L.F.; Jafari, R.; Farzaneh, M. The ice repellency of plasma polymerized hexamethyldisiloxane coating. *Appl. Surf. Sci.* **2013**, 459–463. [CrossRef]
14. Asadollahi, S.; Profili, J.; Farzaneh, M.; Stafford, L. Development of organosilicon-based superhydrophobic coatings through atmospheric pressure plasma polymerization of HMDSO in nitrogen plasma. *Materials* **2019**, *12*, 219. [CrossRef] [PubMed]
15. Pavlos, C.M.; Harke-Bus, R.P.; Ward, K.; Owens, D.E.; Harwood, R.; O'Hare, T.; Ferguson, D.C. Method for Producing Improved Feathers and Improved Feather Thereto. Patent WO 2011/143488, 17 November 2011.
16. Tyczkowski, J.; Kapica, R.; Markiewicz, J.; Malachowski, A.; Malachowski, B. Method for Producing Durable Water-Repellent Layer on the Surface of Natural Down. Patent PL 228924, 19 December 2017.
17. Gao, J.; Yu, W.; Pan, N. Structures and properties of the goose down as a material for thermal insulation. *Text. Res. J.* **2007**, *77*, 617–626. [CrossRef]
18. Bonser, R.H.C.; Dawson, C. The structural mechanical properties of down feathers and biomimicking natural insulation materials. *J. Mater. Sci. Lett.* **1999**, *18*, 1769–1770. [CrossRef]
19. Yildiz, D.; Bozkur, E.U.; Akturks, S.H. Determination of goose feather morphology by using SEM. *J. Anim. Vet. Adv.* **2009**, *8*, 2650–2654. [CrossRef]
20. Stettenheim, P.R. The integumentary morphology of modern birds: An overview. *Amer. Zool.* **2000**, *40*, 461–477. [CrossRef]
21. Webb, D.R.; King, J.R. Effects of wetting on insulation of bird and mammal coats. *J. Therm. Biol.* **1984**, *9*, 189–191. [CrossRef]
22. Bakken, G.S. Wind speed dependence of the overall thermal conductance of fur and feather insulation. *J. Therm. Biol.* **1991**, *16*, 121–126. [CrossRef]
23. Woodcock, A.A.; Steel, N.; Moore, C.B.; Howard, S.J.; Custovic, A.; Denning, D.W. Fungal contamination of bedding. *Allergy* **2006**, *61*, 140–142. [CrossRef]
24. Liu, Y.; Chen, X.; Xin, J.H. Hydrophobic duck feathers and their simulation on textile substrates for water repellent treatment. *Bioinsp. Biomim.* **2008**, *3*, 046007. [CrossRef]
25. Benkovicova, M.; Kisova, Z.; Buckova, M.; Majkova, E.; Siffalovic, P.; Pangallo, D. The antifungal properties of super-hydrophobic nanoparticles and essential oils on different material surfaces. *Coatings* **2019**, *9*, 176. [CrossRef]
26. Shin, S.; Seo, J.; Han, H.; Kang, S.; Kim, H.; Lee, T. Bio-inspired extreme wetting surfaces for biomedical applications. *Materials* **2016**, *9*, 116. [CrossRef]
27. Kim, Y.; Hwang, W. Wettability modified aluminum surface for a potential antifungal surface. *Mater. Lett.* **2015**, *161*, 234–239. [CrossRef]

28. Fan, H.; Guo, Z. Bioinspired surfaces with wettability: Biomolecule adhesion behaviors. *Biomater. Sci.* **2020**, *8*, 1502–1535. [CrossRef] [PubMed]
29. Hydrophobic Shake Test in IDFB Testing Regulations 2020, Version June 2015, Part 18-A. Available online: http://www.cfd.com.cn/upload/contents/2020/08/Testing_Regulations_v202006.pdf (accessed on 10 September 2020).
30. Wang, S.; Jiang, L. Definition of superhydrophobic states. *Adv. Mater.* **2007**, *19*, 3423–3424. [CrossRef]
31. Zhang, X.; Shi, F.; Niu, J.; Jiang, Y.; Wang, Z. Superhydrophobic surfaces: From structural control to functional application. *J. Mater. Chem.* **2008**, *18*, 621–633. [CrossRef]
32. Dorrer, C.; Rühe, J. Some thoughts on superhydrophobic wetting. *Soft Matter* **2009**, *5*, 51–61. [CrossRef]
33. Kurosawa, S.; Choi, B.G.; Park, J.W.; Aizawa, H.; Shim, K.B.; Yamamoto, K. Synthesis and characterization of plasma-polymerized hexamethyldisiloxane films. *Thin Solid Films* **2006**, *506–507*, 176–179. [CrossRef]
34. Grimoldi, E.; Zanini, S.; Siliprandi, R.A.; Riccardi, C. AFM and contact angle investigation of growth and structure of pp-HMDSO thin films. *Eur. Phys. J. D* **2009**, *54*, 165–172. [CrossRef]
35. Kraus, F.; Cruz, S.; Muller, J. Plasmapolymerized silicon organic thin films from HMDSN for capacitive humidity sensors. *Sens. Actuators B Chem.* **2003**, *88*, 300–311. [CrossRef]
36. de Carvalho, A.T.; Carvalho, R.A.M.; da Silva, M.L.P.; Demarquette, N.R. Hydrophobic plasma polymerized hexamethyldisilazane thin films: Characterization and uses. *Mater. Res.* **2006**, *9*, 9–13. [CrossRef]
37. Wang, J.; Chen, H.; Sui, T.; Li, A.; Chen, D. Investigation on hydrophobicity of lotus leaf: Experiment and theory. *Plant. Sci.* **2009**, *176*, 687–695. [CrossRef]
38. Koch, K.; Bhushan, B.; Jung, Y.C.; Barthlott, W. Fabrication of artificial Lotus leaves and significance of hierarchical structure for superhydrophobicity and low adhesion. *Soft Matter* **2009**, *5*, 1386–1393. [CrossRef]
39. Gengenbach, T.R.; Griesser, H.J. Post-deposition ageing reactions differ markedly between plasma polymers deposited from siloxane and silazane monomers. *Polymer* **1999**, *40*, 5079–5094. [CrossRef]
40. Benitez, F.; Martinez, E.; Esteve, J. Improvement of hardness in plasma polymerized hexamethyldisiloxane coatings by silica-like surface modification. *Thin Solid Films* **2000**, *377–378*, 109–114. [CrossRef]
41. Kashiwagi, K.; Yoshida, Y.; Murayama, Y. Hybrid films formed from hexamethyldisiloxane and SiO by plasma process. *Jpn. J. Appl. Phys.* **1991**, *30*, 1803–1807. [CrossRef]
42. Choudhury, A.J.; Chutia, J.; Kakati, H.; Barve, S.A.; Pal, A.R.; Sarma, N.S.; Chowdhury, D.; Patil, D.S. Studies of radiofrequency plasma deposition of hexamethyldisiloxane films and their thermal stability and corrosion resistance behavior. *Vacuum* **2010**, *84*, 1327–1333. [CrossRef]
43. Park, S.Y.; Kim, N.; Kim, U.Y.; Hong, S.I.; Sasabe, H. Plasma polymerization of hexamethyldisilazane. *Polym. J.* **1990**, *22*, 242–249. [CrossRef]
44. Vassallo, E.; Cremona, A.; Ghezzi, F.; Dellera, F.; Laguardia, L.; Ambrosone, G.; Coscia, U. Structural and optical properties of amorphous hydrogenated silicon carbonitride films produced by PECVD. *Appl. Surf. Sci.* **2006**, *252*, 7993–8000. [CrossRef]
45. Kim, M.T.; Lee, J. Characterization of amorphous SiC:H films deposited from hexamethyldisilazane. *Thin Solid Films* **1997**, *303*, 173–179. [CrossRef]
46. Fracassi, F.; Lamendola, R. PECVD of $SiO_XN_YC_ZH_W$ thin films from hexamethyldisilazane containing feed. Investigation on chemical characteristics and aging behavior. *Plasmas Polym.* **1997**, *2*, 25–40. [CrossRef]
47. Li, K.; Gabriel, O.; Meichsner, J. Fourier transform infrared spectroscopy study of molecular structure formation in thin films during hexamethyldisiloxane decomposition in low pressure rf discharge. *J. Phys. D Appl. Phys.* **2004**, *37*, 588–594. [CrossRef]
48. Nouicer, I.; Sahli, S.; Kihel, M.; Ziari, Z. Superhydrophobic surface produced on polyimide and silicon by plasma enhanced chemical vapour deposition from hexamethyldisiloxane precursor. *Int. J. Nanotechnol.* **2015**, *12*, 597–607. [CrossRef]
49. Kazarian, S.G.; Chan, K.L.A. ATR-FTIR spectroscopic imaging: Recent advances and applications to biological systems. *Analyst* **2013**, *138*, 1940–1951. [CrossRef] [PubMed]
50. Alexander, M.R.; Short, R.D.; Jones, F.R.; Michaeli, W.; Blomfield, C.J. A study of $HMDSO/O_2$ plasma deposits using a high-sensitivity and -energy resolution XPS instrument: Curve fitting of the Si 2p core level. *Appl. Surf. Sci.* **1999**, *137*, 179–183. [CrossRef]
51. Chaiwong, C.; Rachtanapun, P.; Sarapirom, S.; Boonyawan, D. Plasma polymerization of hexamethyldisiloxane: Investigation of the effect of carrier gas related to the film properties. *Surf. Coat. Technol.* **2013**, *229*, 12–17. [CrossRef]

52. Grüniger, A.; Bieder, A.; Sonnenfeld, A.; Von Rohr, P.R.; Muller, U.; Hauert, R. Influence of film structure and composition on diffusion barrier performance of SiO_X thin films deposited by PECVD. *Surf. Coat. Technol.* **2006**, *200*, 4564–4571. [CrossRef]
53. Uznanski, P.; Glebocki, B.; Walkiewicz-Pietrzykowska, A.; Zakrzewska, J.; Wrobel, A.M.; Balcerzak, J.; Tyczkowski, J. Surface modification of silicon ocycarbide films produced be remote hydrogen microwave plasma chemical vapour deposition from tetramethyldisiloxane precursor. *Surf. Coat. Technol.* **2019**, *350*, 686–698. [CrossRef]
54. Sahoo, B.; Yoon, K.; Seo, J.; Lee, T. Chemical and physical pathways for fabricating flexible superamphiphobic surfaces with high transparency. *Coatings* **2018**, *8*, 47. [CrossRef]

© 2020 by the authors. Licensee MDPI, Basel, Switzerland. This article is an open access article distributed under the terms and conditions of the Creative Commons Attribution (CC BY) license (http://creativecommons.org/licenses/by/4.0/).

Article

Effect on Silt Capillary Water Absorption upon Addition of Sodium Methyl Silicate (SMS) and Microscopic Mechanism Analysis

Qingwen Ma * and Sihan Liu

School of Water Conservancy Engineering, Zhengzhou University, Zhengzhou 450001, China; m17839941418@163.com
* Correspondence: mqw2008@zzu.edu.cn

Received: 30 June 2020; Accepted: 22 July 2020; Published: 24 July 2020

Abstract: Silt has the characteristics of developed capillary pores and strong water sensitivity, and capillary water is an important factor inducing the erosion and slumping of silt sites. Therefore, in order to suppress the effect of capillary water, this article discusses the improvement effect of sodium methyl silicate (SMS) on silt. The effect was investigated by capillary water rise testing and contact angle measurement, and the inhibition mechanism is discussed from the microscopic view by X-ray diffraction (XRD) testing, X-ray fluorescence (XRF) testing, scanning electron microscope (SEM) testing and mercury intrusion porosimetry (MIP) testing. The results show that SMS can effectively inhibit the rise of capillary water in silt, the maximum height of capillary rise can be reduced to 0 cm when the ratio of SMS (g) to silt (g) increases to 0.5%, and its contact angle is 120.2°. In addition, considering also the XRD, XRF, SEM and MIP test results, it is considered that SMS forms a water-repellent membrane by reacting with water and carbon dioxide, which evenly distribute on the surface of silt particles. The membrane reduces the surface energy and enhances the water repellence of silt, and combines with small particles in the soil, reduces the number of 2.5 µm pores and inhibits the rise of capillary water.

Keywords: sodium methyl silicone; earth site; silt; the height of capillary rise; microscopic mechanism analysis; XRD; XRF; SEM; MIP

1. Introduction

Earth constructions are the remains of human history and culture in a certain environment, which is scientific, historic, artistic and non-renewable. However, a site with silty soil as the main material, because of its special grading characteristics, often has the characteristics of poor stability of the granular skeleton structure, developed capillary pores and strong water sensitivity, so the influence of capillary water on the silt buildings is particularly remarkable [1,2]. Usually, the height of capillary rise in the silt can reach 0.5–1.5 m, or even more than 4 m [3–7]. The moisture content, strength, soluble salt content and microstructure of the soil under the long-term action of the capillary water are all seriously affected, which leads to a decrease in the structural stability and weakening of the foundation [8–11]. Therefore, it is necessary to study methods to control capillary rise in the silt sites.

In recent years, many researchers have improved silt by weakening the capillary rise of the silt. Raw materials such as glutinous rice flour, straw and tung oil were widely applied from the beginning of the Northern and Southern Dynasties (about 420 BC), which obviously changes the impermeability of a building, especially the impermeability of earthen buildings [12,13]; the conventional materials such as cement, fiber-cement, recycled bassanite and the like have also often been applied to improve the soil, and the improved silt is improved not only in strength, but also impermeability, and the rising height of capillary water is also decreased [14–16]. As a new type of material, high-molecular-weight polymer is also often applied to the modification of the silt, in which the detergents, polyalicyclic amine and simplot and the like have been proved to be effective [17–20]. However, earth constructions are not ordinary buildings—they are artistic, scientific, and historic, and cultural relics or cultural heritage is their primary attribute, so it is not suitable to apply the modified methods in the general construction process directly to the earth sites [21].

In order to suppress the capillary rise of the building body, many researchers have used some surfactants in building protection. Potassium methyl silicate showed obvious effects in the tests of inhibiting the hydration swelling and pulp making of mud shale. This prevented water from entering the shale by a hydrophobic membrane formed by organosilicate and the adsorption of potassium ions [22,23]. Sodium methyl silicate (SMS) also performed well in inhibiting the strength and water absorption of concrete. The test results showed that the microstructure of the sample with added sodium methyl silicate was more dense, and insoluble crystals in various shapes were inserted into the crack of concrete; this fully clogged the pores and cracks of the concrete, thus improving the macroscopic properties of concrete, including its waterproofing and impermeability [24–26]. Aiming to address the problem of high moisture absorption of microwave-hardened waterglass sand, some researchers have improved this moisture absorption by using sodium methyl silicate [27,28]. Sodium methyl silicate has also been applied to improve sand in yellow-flooding areas, and it was confirmed that the mechanical properties and the impermeability of the sand can be obviously improved by sodium methyl silicate, as shown by the compaction testing, the strength testing, and permeability testing [29].

Existing research verifies the good effect of SMS on inhibiting the water absorption of concrete, water glass, and other materials, and has a preliminary discussion about the mechanism. However, the use of SMS in the inhibition of capillary water in silt-based sites is rare, and study of the mechanism is insufficient. In order to verify the effect of SMS on capillary water absorption and provide a feasible method for the treatment of capillary water diseases, in this article we investigated the improvement effect of SMS on silt by a capillary water rise experiment and contact angle measurement, and we studied the inhibition mechanism by X-ray diffraction (XRD) testing, X-ray fluorescence (XRF) testing, scanning electron microscope (SEM) testing, and mercury intrusion porosimetry (MIP) testing.

2. Materials and Methods

2.1. Materials and Sample Preparation

Zhengzhou Shang city is the ruins of the capital city the Shang Dynasty (about 1600 BC–1046 BC), located in Guancheng District, Zhengzhou, China. Zhengzhou Shang city is 25 square kilometers and is the largest capital city ruins after the Yin Ruins in the Shang Dynasty. It is of great value for studying the history of the Shang Dynasty and the history of ancient city development. The materials used in the tests were taken from the site of Zhengzhou Shang city, and the soil samples were taken as brownish-yellow silt between 0 and 20 cm from the surface of the ground. Soil samples were taken to the laboratory, and their physical properties were analyzed according to the Highway Geotechnical Test Code (JTG E40-2007) [30]. The results are shown in Table 1. After removing the obvious debris from the soil, the soil was ground and passed through a 2 mm sieve; then we took the required amount of soil samples after screening and dried them in an oven at 105 °C for 12 h to make pretreated dry silt.

Table 1. Basic physical properties of the soil samples.

Soil Type	Density/g·cm^{-3}	Initial Moisture Content/%	Porosity/%	Liquid Limit/%	Plastic Limit/%	Plasticity Index
Silt	1.7	6.8	33.8	23.4	18.1	8.9

According to the test results based on previous preliminary tests and references, the proportions of SMS (g)/dry silt (g) were initially selected as 0%, 0.15%, 0.2%, 0.3%, 0.4%, and 0.5%, and we numbered them sequentially as Sample 0, Sample 1, Sample 2, Sample 3, Sample 4, Sample 5 [22,23,29]. In order to avoid affecting the test results due to different moisture content, the moisture contents of all samples were controlled at 10%, that is, each sample contained 2 kg of dry silt and 0.05 kg of water.

Taking the preparation of Sample 1 as an example, the preparation process can be summarized as follows: first, weigh 2 kg of dry soil, 0.05 kg of water and 3 g of SMS into different containers; secondly, add 0.05 kg of water to the SMS container in small quantities many times, and slowly stir the samples using a glass rod; finally, slowly add the SMS solution to the dry soil, stir the soil thoroughly and place it in a sealed bag for 12 h, so that the solution is evenly distributed throughout the soil. The layered compaction method was used to pour into the capillary water pipe for sample preparation.

2.2. Capillary Water Rise Testing

In order to study the effectiveness of SMS in inhibiting silt capillary water absorption, the above six soil samples were subjected to capillary water rise testing in turn. The specific test equipment and test steps were in accordance with the "Highway Geotechnical Test Code" JTGE40-2007 [30], and we measured the capillary water rising height after the opening of the lock until the rise was stable. Because the maximum height of the test tube was 100 cm, measurements were stopped when they reached 100 cm.

2.3. Contact Angle Measurement

The contact angle is the angle, conventionally measured through the liquid, where a liquid–vapor interface meets a solid surface, and it is written as ω. This value can accurately quantify the degree of soil surface wetting. We observed the water repellency by dropping water first, then used a contact angle instrument to measure the contact angle.

2.4. X-ray Diffraction (XRD) and X-ray Fluorescence (XRF) Testing

XRD (D8 ADVANCE, Brooke, Germany) can determine the main phase of the sample, and XRF can determine the constituent elements of the sample. By combining the results of XRD and XRF, the composition and elements of soil before and after adding SMS can be discussed.

2.5. Scanning Electron Microscopy (SEM) Testing

The principle of a scanning electron microscope (SEM) (Quanta 650, Portland, OR, USA) is to scan a sample with a high-energy electron beam to produce a variety of physical information. By receiving, magnifying and displaying this information, the contact relationship between particles and pores can be reflected directly [31,32]. In order to study the improvement mechanism of SMS from the perspective of microstructure and morphology, it was necessary to observe the microstructure of soil samples before and after adding SMS solution.

2.6. Mercury Intrusion Porosimetry (MIP) Testing

MIP testing can measure the pore size from hundreds of microns to several nanometers, and the equivalent volume of pores can be evaluated by measuring the quantity of mercury entering pores under different external pressure, which can accurately quantify the internal pore morphology of porous materials. MIP has been widely used in different fields [33]. We used MIP in order to better study the mechanism by which methyl sodium silicate inhibits silt capillary water absorption, especially the optimization of the pore distribution.

3. Results and Discussion

3.1. Observations on the Speed of Soil Capillary Water Absorption

The capillary water rising trends of the six soil samples over time are shown in Figure 1. As can be seen from the figure, as time went by, the rising trends of capillary water in the six soil samples were roughly the same. Within the first 8 h of the rise of capillary water, the increasing speed was the fastest in unit time, and then became flat gradually. Sample 0 had the fastest rising speed and took 10 days to reach the maximum rise height. Figure 1 shows that Sample 0 rose rapidly to 100 cm within 120 h, while capillary water in Sample 1 tended to rise steadily and slowly after 48 h until reaching the highest height of 70 cm; Sample 2 tended to rise slowly after 48 h and became stable at about 35 cm; the capillary water in Samples 3 and 4 rose slowly to maximum height and remained stable within 24 h; Sample 5 did not even show any rise of capillary water.

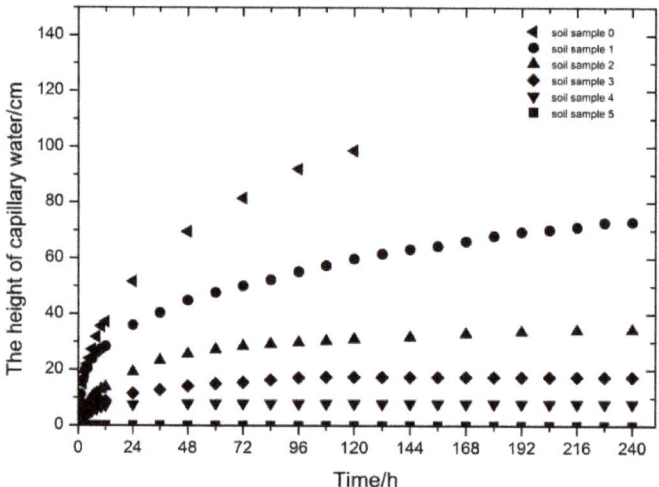

Figure 1. The relationship of the rise height of capillary water with time.

The six test results show that the capillary water rises rapidly in the original soil sample and the rising height can reach more than one meter because of the inferior silt, while the addition of SMS effectively weakens the rise of capillary water in soil. When the ratio of SMS (g) to dry silty soil (g) increases to 0.5%, the capillary rise phenomenon no longer occurs in the test tube.

3.2. Contact Angle Measurement and Analysis

Figure 2 shows the different water droplet forms on the surface of different soil samples. Figure 3 shows contact angle images of Sample 5. As shown in Figure 2, when the droplets were dropped on the surface of the Sample 0, the droplets immediately infiltrated into the soil, and the contact angle was recorded as 0°. Meanwhile the water droplets dropped on the Sample 5 formed an obvious ellipsoid shape, and the contact angle was 120.2°. A material is generally considered to be hydrophobic when the contact angle ω is greater than 90°.

According to the physical and chemical theory of the surface, when the liquid comes into contact with the solid, the liquid will tend to spread. For the liquid phase, there are two main forces: cohesion and adhesion. Cohesion W_c is the attraction between each part of the water molecules, while adhesion W_a is the attraction between the liquid- and solid-phase molecules. The formulas of cohesion and adhesion are Equations (1) and (2), respectively.

$$W_c = 2\sigma_{LG} \tag{1}$$

$$W_a = \sigma_{LG} + \sigma_{SG} - \sigma_{SL} \tag{2}$$

Here, σ_{LG} is the liquid–gas interface attraction; σ_{SL} is the liquid–solid interface attraction; σ_{SG} is the solid–gas interface attraction.

The expression of Gibbs surface free energy is shown in Equation (3).

$$\Delta G/A_S = \sigma_{SL} + \sigma_{LG} - \sigma_{SG} \tag{3}$$

We can substitute Formulas (1) and (2) into Formula (3) to obtain Formula (4).

$$\Delta G/A_S = W_a - W_c \tag{4}$$

According to the second law of thermodynamics, when the reaction is a spontaneous process, we have $\Delta G < 0$. That is, when the cohesion is less than the adhesion, the water will spread and deepen on the surface of the material; when the adhesion is less than the cohesion, water will self-gather on the surface of the material, forming an ellipsoid shape as shown in Figure 2.

Figure 2. Water droplets on the surface of different soil samples: (**a**) Water droplets on the surface of Sample 0; (**b**) Water droplets on the surface of Sample 5.

Figure 3. A contact angle image of Sample 5.

3.3. XRD/XRF Test Results and Analysis

Figure 4 presents the XRD diffraction pattern of Sample 0 and Sample 5, where label (1) indicates the XRD pattern of Sample 0, and label (2) indicates the XRD pattern of Sample 5. Table 2 lists the main elements tested by XRF in the different soils.

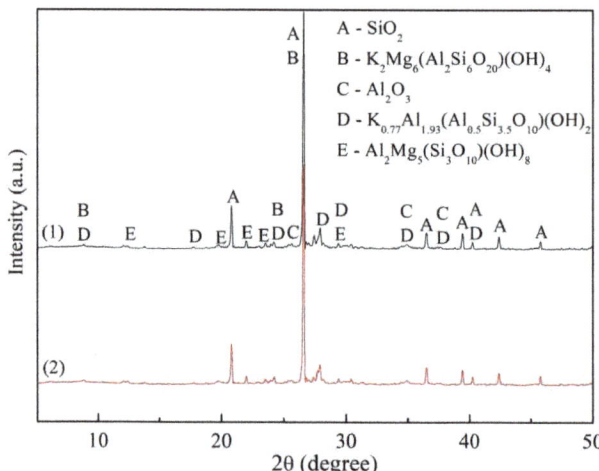

Figure 4. X-ray diffraction (XRD) diffraction patterns of different samples.

Table 2. X-ray fluorescence (XRF) results of Samples 0 and 5.

	Na$_2$O	MgO	Al$_2$O$_3$	SiO$_2$	Cl	K$_2$O	CaO	TiO$_2$	MnO	Fe$_2$O$_3$
Sample 0	1.01	3.44	15.3	62.6	-	3.00	5.04	0.832	0.103	5.25
Sample 5	2.37	3.40	15.0	65.7	0.123	2.81	4.65	0.769	-	4.72

The peak curves in the figures show the following: The main crystal phase of the original soil sample is SiO$_2$, K$_2$Mg$_6$(Al$_2$Si$_6$O$_{20}$)(OH)$_4$, Al$_2$O$_3$, K$_{0.77}$Al$_{1.93}$(Al$_{0.5}$Si$_{3.5}$O$_{10}$)(OH)$_2$, Al$_2$Mg$_5$(Si$_3$O$_{10}$)(OH)$_8$, which means that the soil sample is mainly composed of quartz, mica, montmorillonite, and so on. The X-ray diffraction patterns of the two samples are basically coincident, and the diffraction peaks and diffraction characteristic values only show a slight change. Combined with the results in Table 2, these results indicate that SMS added small amounts of Na and Si to the soil, which are actually Na$_2$CO$_3$ and methyl silicate.

3.4. SEM Testing Results and Analysis

Figures 5 and 6 present SEM images of Samples 0 and 5, respectively. By comparing the scanning electron microscopy pictures of the two, it is obvious that the contact method of soil particles changed significantly after the addition of SMS. The surface of the soil particles before adding SMS was rough, the number of pores was large, and the pore size was big. After adding SMS, the surface edges of the soil particles became relatively smooth and the pore size was greatly reduced.

According to relevant literature, SMS can be decomposed easily by weak acid. When it encounters water and carbon dioxide in the air, it will be decomposed into methyl silicic acid, and then a polymethylsiloxane membrane with waterproof properties will be quickly formed [34]. The chemical process is as follows:

$$2CH_3Si(OH)_2ONa + CO_2 + H_2O \rightarrow 2[CH_3Si(OH)_3] + Na_2CO_3 n[CH_3Si(OH)_3] \rightarrow [CH_3SiO_{3/2}]_n + 3/2H_2O \quad (5)$$

During the preparation of Sample 5, SMS reacted with carbon dioxide and water in the air, and the resulting waterproofing membrane evenly attached to the surface of the soil particles. On the one hand, the water repellency of the membrane makes the soil exhibit strong water repellency. On the other hand, the membrane adheres small particles during the stirring process, changing the way in which the soil particles are combined and narrowing the channels required for water circulation.

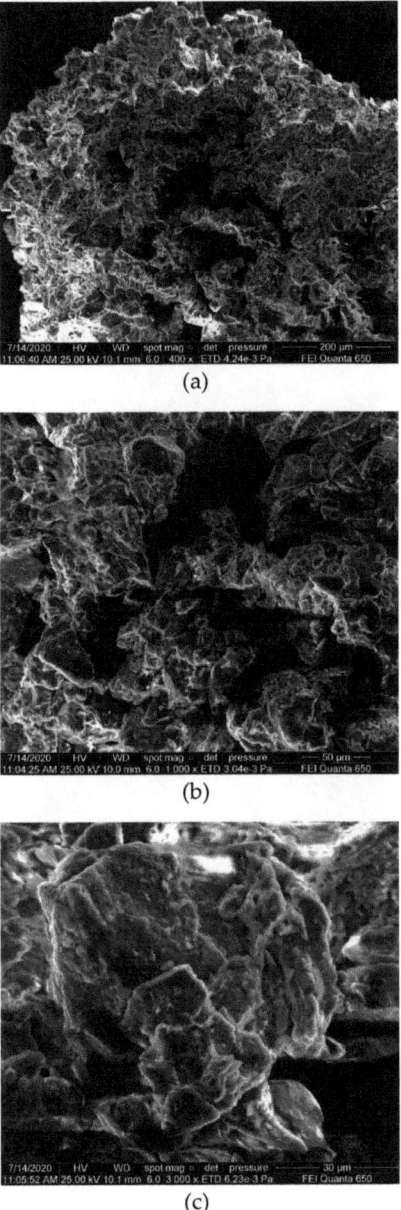

Figure 5. Micromorphology of Sample 0: (**a**) The magnification is 400; (**b**) The magnification is 1000; (**c**) The magnification is 3000.

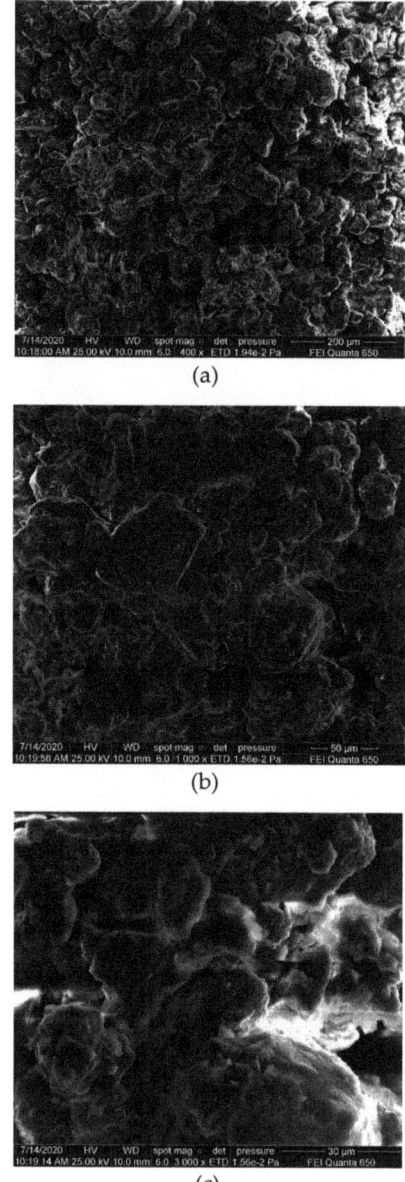

Figure 6. Micromorphology of Sample 5: (**a**) The magnification is 400; (**b**) The magnification is 1000; (**c**) The magnification is 3000.

3.5. MIP Testing Results and Analysis

Cumulative distribution characteristics of pore volume: Mercury injection testing of Samples 0 and 5 was carried out, and the mercury intrusion–extrusion curves of the two samples are given in Figure 7. The mercury intrusion–extrusion curve reflects the changing trend of the total amount of mercury pressed into the pores as the pressure increases. It can be seen that the curves of the two samples are similar in shape, and with increasing pressure, the two cumulative mercury intrusion curves are characterized by a progressive relationship of "sharp rise–slow rise–sharp rise–slow rise". However, with increasing pressure, the mercury injection rate of Sample 0 was significantly lower than that of Sample 5, while when the pressure increased to 20 Pa, the mercury injection rate of Sample 0 increased rapidly and gradually stabilized. Such a rapid change indicated that the number of pores corresponding to the pressure at this time rose substantially. The cumulative total amount of mercury intrusion of Sample 0 was slightly higher than that of Sample 5 in the end, indicating the total porosity of Sample 0 is higher than Sample 5.

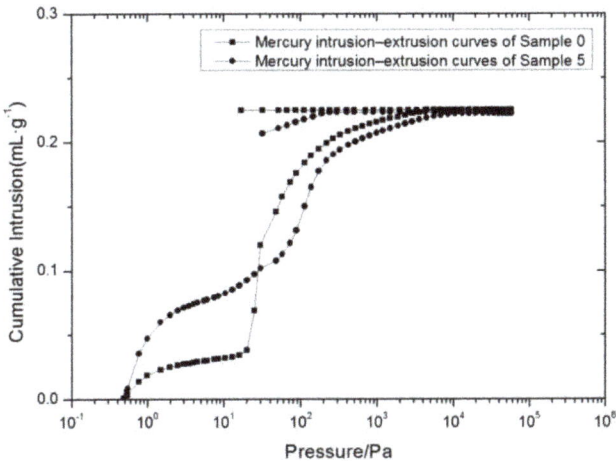

Figure 7. Mercury intrusion-extrusion curve.

Characteristics of pore distribution: during the MIP testing, the pore analysis software can automatically record the mercury intake of each level of pressure, convert them into the corresponding pore diameter, and output the pore size distribution results. After logarithmic processing of the aperture data, the relationship curve of aperture distribution density was obtained. Figure 8 shows that the proportion of small pores and medium pores in the silt was much larger than that of other pores in both soil samples. However, it can be clearly seen that the pores with a diameter of 1–10 μm of the Sample 0 have a sudden and large increase, which is consistent with the results in Figure 7. At the same time, compared with Sample 0, the pore size distribution of Sample 5 was relatively uniform; the percentage of large pores increased, the percentage of medium pores decreased significantly, the percentage of small pores increased, and the percentage of micropores and ultrafine pores remained almost the same.

Professor Shear gives the following pore divisions [35]: large pores (>10 μm), mainly intergranular pores; medium pores (2.5–10 μm), mainly intraparticle pores; small pores (0.4–2.5 μm), mainly intergranular and partially intragranular pores; micropore (0.03–0.4 μm), belonging to intergranular pores; ultrafine pores (<0.03 μm), mainly intraparticle pores.

The absorption of capillary water mainly depends on the medium pores and large pores inside the material; that is, pores with a pore diameter of more than 2.5 μm. Table 3 shows that the pore ratio of Sample 5 increased slightly and the specific surface area increased greatly, which indicates that the medium pores decreased and the small pores increased in number in Sample 5. This can reasonably explain the decrease of the rising height of capillary water in Sample 5.

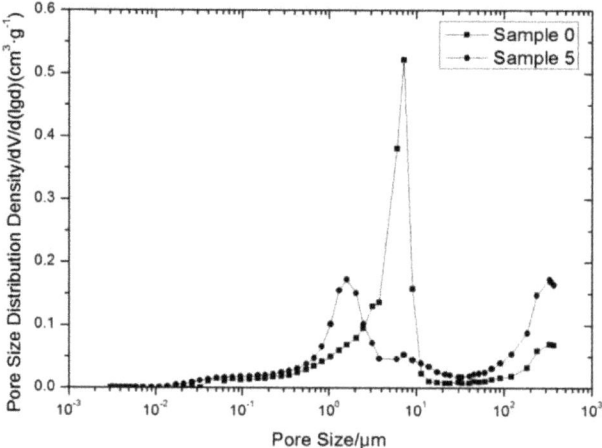

Figure 8. Density curve of pore size distribution.

Table 3. Intrusion data summary.

	Total Intrusion Volume (mL/g)	Total Pore Area (m^2/g)	Porosity (%)
Soil sample 0	0.2253	0.878	33.8093
Soil sample 5	0.2224	1.544	36.3578

4. Conclusions

As the SMS content increased, the maximum height of capillary rise gradually decreased from 121.2 cm to 0 cm, verifying the good effect of SMS in inhibiting the capillary water rise. Contact angle exceeded 120°, proving that the soil has good water repellency. From the combined XRD results, SEM images, and MIP results, it can be seen that after entering the soil, SMS solution was evenly dispersed with water and penetrated into the porous surfaces to form a waterproof and breathable polymethylsiloxane membrane on the surface of silty particles. The membrane had two effects on silt particles: it enclosed silt particles, and it bound adjacent silt particles together.

SMS can effectively suppress the absorption of capillary water without changing the appearance of the soil or reacting with the soil; therefore, the treatment of capillary water disease in silty sites can be achieved by using SMS added to silt as a repair material or by applying SMS solution on the site surface. The results of this study can provide an engineering basis for the treatment of capillary water disease in silty soil sites, given the insufficient existing data on the treatment of capillary water diseases.

Author Contributions: Conceptualization, Q.M.; methodology, Q.M.; validation, Q.M. and S.L.; formal analysis, Q.M. and S.L.; investigation, Q.M. and S.L.; resources, Q.M.; data curation, S.L.; writing—original draft preparation, S.L.; writing—review and editing, Q.M., and L.S; visualization, S.L.; supervision, Q.M.; project administration, Q.M.; funding acquisition, Q.M. All authors have read and agreed to the published version of the manuscript.

Funding: This research was funded by Science and Technology Development Plan of Henan Province in 2018, grant No. 182102310900.

Conflicts of Interest: The authors declare no conflict of interest.

References

1. Liu, C.H. Study on the Water Movement of Silt and the Seepage Characteristics of the City Wall Considering Clay content. Master's Thesis, Zhongyuan Institute of Technology, Henan, China, April 2019.
2. Li, L. Study on the Freeze-Thaw Cycle Effect and Micromechanism of the Strength and Disintegration of the Silt at the Site. Master's Thesis, Zhongyuan Institute of Technology, Henan, China, April 2019.

3. Yan, C.G.; Wan, Q.; Xu, Y.; Xie, Y.; Yin, P. Experimental study of barrier effect on moisture movement and mechanical behaviors of loess soil. *Eng. Geol.* **2018**, *240*, 1–9. [CrossRef]
4. He, F.; Pan, Y.; Tan, L.; Zhang, Z.; Li, P.; Liu, J.; Song, X. Study of the water transportation characteristics of marsh saline soil in the Yellow River Delta. *Sci. Total Environ.* **2017**, *574*, 716–723. [CrossRef]
5. Zhang, Y.C.; Zhang, X.F.; Leng, Y.F. The research on stable rising height and harmful rising height of capillary water. In Proceedings of the 2011 International Conference on Transportation and Mechanical & Electrical Engineering (TMEE), Changchun, China, 16 December 2011; pp. 2190–2197.
6. Dong, B.; Zhang, X.F.; Li, X.; Zhang, D.Q. Comprehensive test research of capillary water rising height. *Chin. J. Geotech. Eng.* **2008**, *10*, 1569–1574.
7. Spennemann, D.H.R. The creeping disaster: Dryland and urban salinity and its impact on heritage. *Cult. Resour. Prot. Emerg. Prep.* **2001**, *24*, 22–26.
8. Ren, K.B.; Wang, B.; Li, X.M.; Yin, S. Strength characteristics and pore distribution characteristics of soil ruins under the action of capillary water dry-wet cycle. *Rock Soil Mech.* **2019**, *40*, 962–970.
9. Zhu, D.Y.; Guan, Y.H. The effect of capillary water on the stability of silt subgrade. *J. Shandong Univ.* **2012**, *42*, 9398.
10. Walker, P. Terra 2003. In Proceedings of the 9th International Conference on the Study and Conservation of Earthen Architecture, Terra, Iran, 29 November 2003; GCI: Los Angeles, CA, USA, 2003.
11. Rainer, L.H. Water, wind, salt, biological, environmental, deterioration/pathology. In *Terra Literature Review*; GCI: Los Angeles, CA, USA, 2003.
12. Cui, C. Disease Analysis of a City Wall Soil Site and the Stability of Modified Loess after Restoration. Master's Thesis, Xi'an University of Technology, Shanxi, China, June 2019.
13. Zhang, H.Y.; Zhu, S.B.; Li, M.; Zhang, X.C. Water repellency of monument soil treated by tung oil. *Geotech. Geol. Eng.* **2016**, *34*, 205–216. [CrossRef]
14. Yuan, Y.; Zhao, L.; Li, W.; Cao, R. Research on silty soil capillary water rising in yellow river flooded area of eastern henan. *J. Highw. Transp. Res. Dev.* **2016**, *10*, 40–46. [CrossRef]
15. Ahmed, A. Recycled bassanite for enhancing the stability of poor subgrades clay soil in road construction projects. *Constr. Build. Mater.* **2013**, *48*, 151–159. [CrossRef]
16. Ahmed, A.; Ugai, K.; Kamei, T. Investigation of recycled gypsum in conjunction with waste plastic trays for ground improvement. *Constr. Build. Mater.* **2011**, *25*, 208–217. [CrossRef]
17. Shafran, A.W.; Gross, A.; Ronen, Z.; Weisbrod, N.; Adar, E. Effects of surfactants originating from reuse of greywater on capillary rise in the soil. *Water Sci. Technol.* **2005**, *52*, 157–166. [CrossRef] [PubMed]
18. Dong, J.M.; Xu, H.Z.; Zhu, D.H.; Zhu, H. Experimental study on polymer-modified silt under different water environments. *Chin. J. Geotech. Eng.* **2013**, *35*, 1316–1322.
19. Mobbs, T.L.; Peters, R.T.; Davenport, J.R.; Evans, M.A.; Wu, J.Q. Effects of four soil surfactants on four soil-water properties in sand and silt loam. *J. Soil Water Conserv.* **2012**, *67*, 275–283. [CrossRef]
20. Abu-Zreig, M.; Rudra, R.P.; Dickinson, W.T. Effect of application of surfactants on hydraulic properties of soils. *Biosyst. Eng.* **2003**, *84*, 363–372. [CrossRef]
21. Kong, D.; Wan, R.; Chen, J.; Jing, Y.; Huang, W.; Wang, Y. The study on engineering characteristics and compression mechanisms of typical historical earthen site soil. *Constr. Build. Mater.* **2019**, *213*, 386–403. [CrossRef]
22. Han, W.C.; Li, Y.; Tan, X.F.; Guo, M.Y.; Xu, H.W. Methyl silicate inhibits clay hydration and its mechanism. *Prospect. Eng. (Rock & Soil Drill. Eng.)* **2018**, *45*, 19–23.
23. Jiang, G.C.; Wang, J.S.; Xuan, Y. Performance evaluation and action mechanism of potassium methyl silicate shale inhibitor. *Sci. Technol. Eng.* **2014**, *14*, 6–10.
24. Yang, H.T.; Yu, J.J. Research on the influence of active components on the properties of concrete. *New Build. Mater.* **2019**, *46*, 131–133.
25. Guo, Z.; Zhu, Q.; Liu, C.; Xing, Z. Preparation of Ca-Al-Fe deicing salt and modified with sodium methyl silicate for reducing the influence of concrete structure. *Constr. Build. Mater.* **2018**, *172*, 263–271. [CrossRef]
26. Liu, G. Experimental Study on Adding Chloride Ion Transmission Blocking Materials to Cement Concrete. Master's Thesis, Chang'an University, Shanxi, China, May 2017.
27. Li, X.J. Microwave Hardened Water Glass Sand Composite Hardening Process and Anti-Hygroscopicity. Master's Thesis, Huazhong University of Science and Technology, Hubei, China, January 2013.

28. Wang, H.F. Key Technical Foundations of Microwave Hardened Water Glass Sand Application. Ph.D. Thesis, Huazhong University of Science and Technology, Hubei, China, January 2012.
29. Yuan, Y.Q.; Jia, M.; Li, W. Experimental study of yellow silty soil stabilized by sodium methyl silicate. *China Sci. Technol. Pap.* **2017**, *12*, 831–844.
30. JTGE40-2007. *Highway Geotechnical Test Regulations [S]*; People's Communications Press: Beijing, China, 2007.
31. Zhang, W.P.; Sun, Y.F.; Tong, W.W.; Song, Y.P.; Dong, L.F.; Liu, X.Y. An analytical method for studying the distribution characteristics of soil particles and pores based on sem images. *Adv. Mar. Sci.* **2018**, *36*, 605–613.
32. Liu, Y.J.; Wu, J.S.H.; Xie, Z.H. Experimental study on microstructure characteristics of soft soil based on NMR and SEM. *J. Guangdong Univ. Technol.* **2018**, *35*, 49–56.
33. Wu, S.; Yang, J.; Yang, R.; Zhu, J.; Liu, S.; Wang, C. Investigation of microscopic air void structure of anti-freezing asphalt pavement with X-ray CT and MIP. *Constr. Build. Mater.* **2018**, *178*, 473–483. [CrossRef]
34. Lu, F.; Nie, J.T. Application research of sodium methyl silicate (organic silicone water repellent). *Jiangxi Chem. Ind.* **1995**, *2*, 31–34.
35. Shear, D.L.; Olsen, H.W.; Nelson, K.R. Effects of desiccation on the hydraulic conductivity versus void ratio relationship for a natural clay. In *Transportation Research Record*; National Academy Press: Washington, DC, USA, 1993; pp. 1365–1370.

© 2020 by the authors. Licensee MDPI, Basel, Switzerland. This article is an open access article distributed under the terms and conditions of the Creative Commons Attribution (CC BY) license (http://creativecommons.org/licenses/by/4.0/).

Communication

Superhydrophobic Coatings Based on Siloxane Resin and Calcium Hydroxide Nanoparticles for Marble Protection

Aikaterini Chatzigrigoriou [1], Ioannis Karapanagiotis [2,*] and Ioannis Poulios [1]

[1] Department of Chemistry, Aristotle University of Thessaloniki, 54124 Thessaloniki, Greece; katerinaxatzigrigoriou@yahoo.gr (A.C.); poulios@chem.auth.gr (I.P.)
[2] Department of Management and Conservation of Ecclesiastical Cultural Heritage Objects, University Ecclesiastical Academy of Thessaloniki, 54250 Thessaloniki, Greece
* Correspondence: y.karapanagiotis@aeath.gr

Received: 12 March 2020; Accepted: 27 March 2020; Published: 1 April 2020

Abstract: Calcium hydroxide ($Ca(OH)_2$) nanoparticles are produced following an easy, ion exchange process. The produced nanoparticles are characterized using transmission electron microscopy (TEM) and Fourier- transform infrared spectroscopy (FTIR) and are then dispersed in an aqueous emulsion of silanes/siloxanes. The dispersions are sprayed on marble and the surface structures of the deposited coatings are revealed using scanning electron microscopy (SEM). By adjusting the nanoparticle concentration, the coated marble obtains superhydrophobic and water repellent properties, as evidenced by the high static contact angles of water drops (> 150°) and the low sliding angles (< 10°). Because $Ca(OH)_2$ is chemically compatible with limestone-like rocks, which are the most common stones found in buildings and objects of the cultural heritage, the produced composite coatings have the potential to be used for conservation purposes. For comparison, the wetting properties of another superhydrophobic and water repellent coating composed of the same siloxane material and silica (SiO_2) nanoparticles, which were commonly used in several previously published reports, were investigated. The suggested siloxane+$Ca(OH)_2$ composite coating offers good protection against water penetration by capillarity and has a small effect on the aesthetic appearance of marble, according to colorimetric measurements.

Keywords: superhydrophobic; water repellency; calcium hydroxide; siloxane; marble; cultural heritage; conservation

---

1. Introduction

Recent advances in coating materials may offer novel routes for effective and sustainable protection and preservation of natural stone used in cultural heritage [1]. For example, superhydrophobic and water repellent coating materials can offer protection against the deteriorative effects of rainwater, as they can reduce the penetration of atmospheric liquid water into the pore network of natural stone.

The static contact angle (CA) of a water drop on a superhydrophobic surface is CA > 150°, whereas the sliding angle (SA) of a water drop on a water repellent surface is SA < 10°. Superhydrophobicity is usually (e.g., lotus leaf [2]), but not always (e.g., rose petal [3]), accompanied by water repellency. Therefore, both CA and SA are important to actually characterize the wetting properties of a material.

Research results of the last fifteen years have shown that an easy and effective strategy to produce superhydrophobic and water repellent coatings is the integration (addition) of nanoparticles into a low surface energy polymer matrix [4,5]. Nanoparticles enhance surface roughness, which is the key parameter to achieve extreme wetting properties. This method was suggested to produce polymer+nanoparticle composite coatings for the protection of natural stone in 2007 [6] and was established through relevant detailed studies in 2009 [7,8]. In these early works, silica (SiO_2) nanoparticles were used as additives to roughen the surface of siloxane, acrylic, and perfluorinated polymer coatings [6–8]. Since then, SiO_2 nanoparticles have become the standard additives for the production of superhydrophobic polymer+nanoparticle composite coatings for natural stone protection [9–17]. Other nanoparticles, selected for the same purpose, are aluminum oxide (Al_2O_3) [8] and tin oxide (SnO_2) [8], as well as photo-catalytic and biocidal nanomaterials, such as titanium oxide (TiO_2) [8,13,18–21], zinc oxide (ZnO) [12,19], and silver (Ag) [22].

The goal of the present short study is to produce a superhydrophobic and water repellent siloxane-based composite coating by adding calcium hydroxide ($Ca(OH)_2$) nanoparticles. Unlike the nanoparticles described above and those used in the past, $Ca(OH)_2$ is chemically compatible with limestone and limestone-like rocks (marble, travertine), which are undoubtfully the most common stones that have been used in the past [1]. Nanoparticles of $Ca(OH)_2$ are produced, characterized, mixed with a siloxane-based precursor in various concentrations, and sprayed onto marble specimens to evaluate the wettabilities of the produced composite coatings. For comparative purposes, other composite coatings were prepared using the standard SiO_2 nanoparticles that were utilized in the past to achieve extreme wetting properties [6–17].

2. Materials and Methods

2.1. Materials

Materials used for the production of the $Ca(OH)_2$ nanoparticles were calcium chloride dihydrate (≥ 99%, $CaCl_2·2H_2O$), which was obtained from Chem-Lab (Zedelgem Belgium) and an anion exchange resin, Dowex Monosphere-550A-OH (Delfgauw, The Netherlands). Silica (SiO_2) nanoparticles of 7 nm in mean diameter were purchased from Sigma-Aldrich (St. Louis, MO, USA). Our group has used these purchased SiO_2 nanoparticles in the past to produce superhydrophobic coatings in several investigations, as described in a review book chapter [4]. A water based emulsion composed of amino-modified silanes and fluoro-modified siloxanes was used for the preparation of the dispersions. The ratio of (silanes+siloxanes):water in the emulsion was 1:6 (v/v). Finally, specimens of dolomitic marble, from Thassos, Greece [7], were used in the study.

2.2. Synthesis and Characterization of $Ca(OH)_2$ Nanoparticles

$Ca(OH)_2$ nanoparticles were produced according to a recently devised method, which is based on an ion exchange process between an anionic resin and a calcium chloride aqueous solution at room temperature (r.t.) [23]. In particular, an aqueous solution containing 0.1 M of $CaCl_2•2H_2O$ was prepared and mixed with 75 mL of anion exchange resin, at r.t. and under moderate stirring for 15 min. The resin was then separated from the suspension using a sieve with mesh size of 200 μm. The separated suspension was mixed with 75 ml of fresh resin at r.t. and under moderate stirring for 30 min. The resin was removed again by sieving and the suspension was subjected to centrifugation for 15 min at 6000 rpm. The precipitated nanoparticles were dried in a vacuum at 50 °C for 12 h.

The produced $Ca(OH)_2$ nanoparticles were characterized using transmission electron microscopy (TEM; JEOL, 2000FX, Tokyo, Japan) and Fourier-transform infrared spectroscopy (FTIR), which was employed using a Spectrum Spotlight 400 PerkinElmer spectrometer (Waltham, MA, USA).

2.3. Production and Characterisation of Siloxane+Nanoparticle Coatings

The produced $Ca(OH)_2$ nanoparticles were dispersed in the silane/siloxane emulsion in various concentrations. The dispersions were stirred mechanically and sprayed onto marble specimens using an airbrush system with a nozzle of 660 μm in diameter (Paasche Airbrush, Chicago, IL, USA). Another set of coated marbles were prepared using the silane/siloxane emulsion and SiO_2 nanoparticles. Moreover, pure silane/siloxane emulsion (without nanoparticles) was also sprayed onto marble. The morphologies of the deposited coatings were investigated using scanning electron microscopy (SEM; JEOL, JSM-6510, Tokyo, Japan).

Static contact angle (CA) and sliding angle (SA) were measured using an optical tensiometer apparatus (Attension Theta, Gothenburg, Sweden). For the measurements of the SAs, the tilt rate was adjusted to 1°/s. The reported angles are averages of five measurements. Variations for the measurements of CAs are provided, and the SAs varied within ± 1°.

The measurements of water capillary absorption were performed by the gravimetric sorption technique. Dried weighted coated and uncoated marble blocks were placed on a filter paper pad (Whatman paper, No. 4, Little Chalfont, UK) partially immersed in distilled water. Samples were weighted periodically for a period of 15 h in total to measure the amount of water absorbed by the specimens. Finally, colorimetric measurements were carried out using a Miniscan XE Plus spectrophotometer (HunterLab, Reston, VA, USA). The reported results are averages of three measurements, and variations are reported.

3. Results

The produced nanoparticles were characterized using TEM and FTIR, as shown in Figure 1. According to the TEM image, the sizes of the produced particles were lower than 100 nm, indicating that particles at the nanometer scale were successfully produced. The FTIR spectrum shows characteristic peaks which lead to the identifications of the carbonate ($CaCO_3$) and hydroxide ($Ca(OH)_2$) compounds of Ca [24,25]. In particular, the bands at 1444, 877, and 714 cm^{-1} correspond to the three different elongation modes of C–O bonds, while the bands at 2983, 2875, and 2513 cm^{-1} are harmonic vibration of these elongation modes. The thin band at 1795 cm^{-1} is associated to the carbonate C=O bonds. The strong band at 3643 cm^{-1} is related to the O–H bonds from hydroxides [24,25].

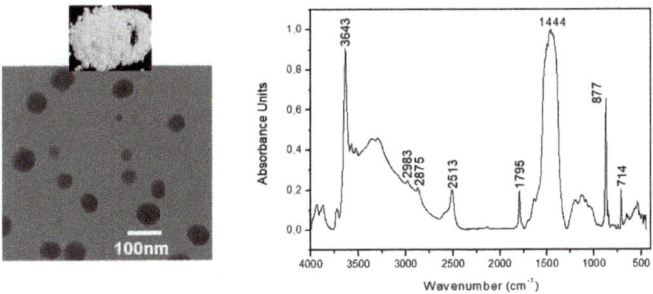

Figure 1. Photograph, TEM image, and FTIR spectrum of the produced nanoparticles.

The surfaces of the siloxane+$Ca(OH)_2$ coatings were characterized using SEM. Indicative images are provided in Figure 2. Adding nanoparticles to the coating's composition results in the formation of surface structures-protrusions that consist of siloxane material mixed with particle agglomerations. As the nanoparticle concentration increases, the surface protrusions become denser, thus promoting surface roughness. The latter is responsible for the extreme wetting properties, which are discussed in the next paragraphs. The scenario revealed in Figure 2 for the siloxane+$Ca(OH)_2$ coatings follows the results reported previously for the effect of SiO_2 nanoparticles on the surface structure of siloxane-based composite coatings [6–17]. Surface structures reported for siloxane+SiO_2 coatings on marble [6–17] are similar to those shown in Figure 2 for the siloxane+$Ca(OH)_2$ coatings.

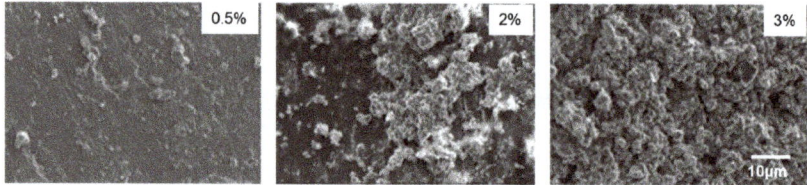

Figure 2. Evolution of the surface structures of the siloxane+Ca(OH)$_2$ coatings. The concentrations (w/w) of the Ca(OH)$_2$ nanoparticles in the dispersions that were deposited on marble are shown in the upper right corner of each image.

Figure 3 shows the results of the CA and SA measurements of water drops placed on the surfaces of the composite coatings on marble. Two sets of data are included corresponding to the coatings that were prepared using the produced Ca(OH)$_2$ and the purchased SiO$_2$ nanoparticles. According to Figure 3a, the CA of water drops on the surface of pure siloxane (without nanoparticles) is 113° ± 3°, indicating that the application of the water-based emulsion results in the formation of a coating, which shows hydrophobicity [5]. The results in Figure 3a show that CA increases with nanoparticle concentration and eventually becomes very large. Superhydrophobicity (CA > 150°) is evidenced for coatings that were prepared using concentrations of Ca(OH)$_2$ > 1.5 % w/w and SiO$_2$ > 1 % w/w. Overall, siloxane+SiO$_2$ coatings gave larger CAs, compared to the results reported in Figure 3a for the siloxane+Ca(OH)$_2$ coatings. This difference is within the experimental error for the coatings, which were prepared using the maximum nanoparticle concentration tested herein (3 % w/w), i.e., the error bars of CAs for the two coatings prepared with SiO$_2$ and Ca(OH)$_2$ nanoparticles overlap. The results in Figure 3a are in agreement with previously published reports that revealed the cross influence effects of particle size and concentration on the wettability of siloxane+nanoparticle composite films. In particular, it was shown that hydrophobicity is enhanced with (i) nanoparticle concentration up to a saturation point [6–9] and (ii) decreasing particle size [26,27]. In the results of Figure 3a, it is seen that CA increases with nanoparticle concentration reaching a plateau (saturation), which is clearer in the case of the SiO$_2$ nanoparticles. Moreover, larger CAs are obtained with the siloxane+SiO$_2$ coatings, as the SiO$_2$ nanoparticles are one to two orders of magnitude smaller than the Ca(OH)$_2$ nanoparticles.

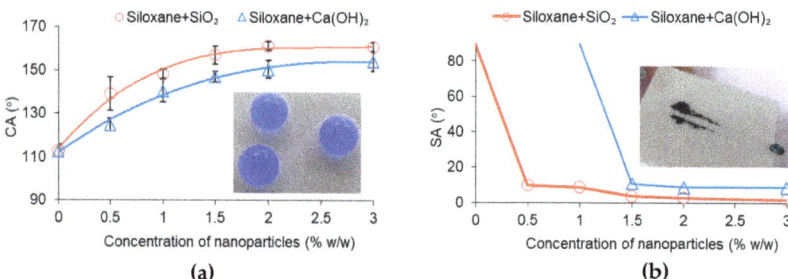

Figure 3. (**a**) Static contact angle (CA) and (**b**) sliding angle (SA) vs the nanoparticle concentration. Data points were (**a**) fitted with polynomial functions and (**b**) connected with lines to guide the eye. Coatings prepared using < 0.5 % w/w SiO$_2$ and < 1.5 % w/w Ca(OH)$_2$ were pinned on the surfaces; therefore, they correspond to a theoretical SA of 90°. Photographs showing (**a**) resting drops and (**b**) the self-cleaning process were taken for marble specimens that were coated with superhydrophobic and water repellent composite coatings. The latter were prepared using siloxane and 3 % w/w Ca(OH)$_2$ nanoparticles.

Figure 3b shows the measurements of SA of water drops on coated marble specimens. Coatings that were prepared using < 0.5 % w/w SiO$_2$ and < 1.5 % w/w Ca(OH)$_2$ showed water adhesion, as the water drops were pinned on these surfaces even when they were tilted by 90°. Therefore, it was not possible to actually measure SAs on these surfaces which correspond to a theoretical SA = 90°. The SA of water drops on the siloxane+SiO$_2$ surfaces decreased rapidly with nanoparticle concentration and eventually became very small (SA < 10°) and stable when the nanoparticle concentration became > 1 % w/w. For the coatings that were prepared using the bigger Ca(OH)$_2$ nanoparticles, higher nanoparticle concentration (> 1.5 % w/w) had to be used to achieve water repellency (SA < 10°).

The superhydrophobic and water repellent performance of the siloxane + 3 % w/w Ca(OH)$_2$ composite coating is demonstrated in the photographs of Figure 3a,b. Resting drops and the self-cleaning process on coated marble specimens are shown in the two photographs. Consequently, the wetting properties of the composite coating mimic those of the lotus leaf surface [2].

The interaction of the siloxane + 3 % w/w Ca(OH)$_2$ coating with water was further tested by performing measurements of water capillary absorption. For comparison, uncoated marble blocks and blocks coated by pure siloxane, without nanoparticles, were included in the study. The amount of water absorbed by the specimen per unit area (Q_i) was calculated as follows:

$$Q_i = \left(\frac{w_i - w_o}{A}\right) \times 100, \qquad (1)$$

where w_i is the weight of the sample after being in contact with water for time t_i, w_o is the initial weight of the sample prior to the test, and A is the sample's area which was in contact with water during the test. The calculated Q_i values were plotted as a function of time t_i, and the results are presented in Figure 4.

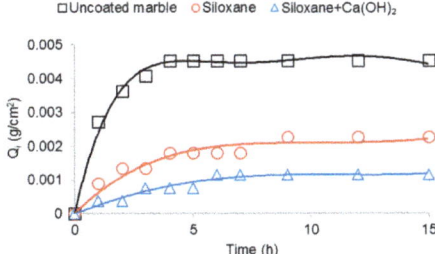

Figure 4. Amounts of absorbed water per unit area (Q_i) as a function of treatment time t_i, for three samples: (i) uncoated marble; (ii) marble coated by pure siloxane; and (iii) marble coated by a composite coating, which was prepared using siloxane and 3% w/w Ca(OH)$_2$ nanoparticles (siloxane+Ca(OH)$_2$). Data points were fitted with polynomial functions to guide the eye.

The results of Figure 4 show that the specimens became saturated in absorbed water. This is evidenced by the recorded plateaus of the three Q_i - t_i curves. The amounts of absorbed water follow the order: uncoated sample -> sample coated by siloxane -> sample coated by siloxane+ Ca(OH)$_2$, with the latter being the sample that absorbed the least amount of water at each specific t_i. Specifically, by taking into consideration the last three (t_i = 9, 12, and 15 h) measurements of Q_i for each curve, which clearly correspond to the plateaus of the curves, average-maximum values of Q_i were calculated as follows: 0.0045 g/cm^2 for the uncoated sample, 0.0023 g/cm^2 for the sample coated by siloxane, and 0.0012 g/cm^2 for the sample coated by siloxane+Ca(OH)$_2$. The hydrophobic siloxane coating offers protection against the capillary absorption of water. The protection, however, is enhanced when the superhydrophobic and water repellent siloxane+Ca(OH)$_2$ coating is applied on the marble surface.

For the maxima amounts of absorbed water, corresponding to the plateaus of the curves in Figure 4, the relative reduction of water absorption by capillarity (RC%) was calculated using the following Equation:

$$RC\% = \left(\frac{Q_u - Q_c}{Q_u}\right) \times 100, \qquad (2)$$

where Q_u and Q_c are the maxima Q_i measured for the uncoated and coated marble specimens, respectively. An ideal coating should correspond to RC% = 100, as it should eliminate the penetration of liquid water into the pore network of the stone. Using Equation (2), it is calculated that the application of the siloxane coating on marble results in a reduction of the amount of absorbed water by 49%. The RC% increased to 73 when the superhydrophobic and water repellent siloxane+Ca(OH)$_2$ coating was applied on marble.

Finally, the optical effects of the siloxane and siloxane+Ca(OH)$_2$ coatings on marble were evaluated through colorimetric measurements. The global color differences (ΔE*) of marble, induced upon coating application, was derived from:

$$\Delta E^* = \sqrt{(L_c^* - L_u^*)^2 + (a_c^* - a_u^*)^2 + (b_c^* - b_u^*)^2}, \tag{3}$$

where L*, a*, and b* are the components of the CIE 1976 scale, respectively. The "c" and "u" subscript characters indicate the coated and uncoated specimens, respectively. The results of the L*, a*, and b* measurements are summarized in Table 1. Using Equation (3) and the values of Table 1, it is calculated that ΔE* = 0.36 ± 0.04 for the marble specimen that was coated with siloxane and ΔE* = 3.76 ± 0.03 for the marble specimen that was coated with siloxane+Ca(OH)$_2$. According to the literature, color variations which correspond to ΔE* < 3 are insignificant as they are not perceived by human eye [28–30]; the accepted level for conservation purposes is ΔE* < 5 [28]. Consequently, the results, which are reported in Table 1, suggest that the siloxane material (ΔE* = 0.36) has a negligible effect on the color of the marble. However, when Ca(OH)$_2$ nanoparticles are added in the coating, then the treatment of the marble is accompanied by a noticeable visual effect (ΔE* = 3.76) which, however, is not very far away from the human perception threshold value and clearly below the accepted level for conservation purposes.

Table 1. Color coordinates measured for uncoated marble and marble specimens coated with siloxane and siloxane+Ca(OH)$_2$. The composite coating was prepared using 3 % w/w Ca(OH)$_2$ nanoparticles.

	Uncoated	Siloxane	Siloxane+Ca(OH)$_2$
L*	91.67 ± 0.03	91.34 ± 0.05	95.41 ± 0.01
a*	−0.24 ± 0.01	−0.29 ± 0.01	−0.06 ± 0.01
b*	3.45 ± 0.01	3.57 ± 0.02	3.16 ± 0.02

4. Conclusions

The major finding from this work is that superhydrophobic and water repellent coatings can be produced using dispersions of Ca(OH)$_2$ nanoparticles in water-based silane/siloxane emulsions. Ca(OH)$_2$ is chemically compatible with limestone and limestone-like rocks (marble, travertine), which are undoubtfully the most common stones found in buildings and objects of the cultural heritage. Siloxane-based materials are commonly used for consolidation purposes. Therefore, the siloxane+Ca(OH)$_2$ composite coatings, which are formed by spraying the aforementioned dispersions onto the stone substrate, have the potential to be used for conservation purposes.

The Ca(OH)$_2$ nanoparticles (Figure 1) were synthesized following an easy, ion exchange process, and the extreme wetting properties, which were achieved on the surface of the composite coatings (Figure 2), were evidenced by the high CA > 150° and the low SA < 10° (Figure 3) of water drops. Composite coatings offered good protection against water penetration by capillarity (Figure 4) and had a small effect on the color of marble (Table 1).

Further and more detailed studies should be carried out in the future to investigate the durability of the siloxane+Ca(OH)$_2$ coatings and their effects on the breathability of marble.

Author Contributions: Conceptualization, I.K.; investigation, A.C.; data curation, A.C. and I.K.; writing—original draft preparation, I.K.; writing—review and editing, I.K. and I.P.; project administration, I.P.; All authors have read and agreed to the published version of the manuscript.

Funding: This research received no external funding.

Acknowledgments: The authors would like to thank Shrief Eissa for his assistance with the TEM studies, Dimitrios Lampakis for his contribution to the FTIR measurements and Dimitra Aslanidou for her assistance in the production of the nanoparticles.

Conflicts of Interest: The authors declare no conflict of interest.

References

1. Artesani, A.; Di Turo, F.; Zucchelli, M.; Traviglia, A. Recent advances in protective coatings for cultural heritage—An overview. *Coatings* **2020**, *10*, 217. [CrossRef]
2. Barthlott, W.; Neinhuis, C. Purity of the sacred lotus, or escape from contamination in biological surfaces. *Planta* **1997**, *202*, 1–8. [CrossRef]
3. Feng, L.; Zhang, Y.; Xi, J.; Zhu, Y.; Wang, N.; Xia, F.; Jiang, L. Petal effect: A superhydrophobic state with high adhesive force. *Langmuir* **2008**, *24*, 4114–4119. [CrossRef]
4. Karapanagiotis, I.; Manoudis, P. Superhydrophobic and water repellent polymer-nanoparticle composite films. In *Industrial Applications for Intelligent Polymers and Coatings*; Hosseini, M., Makhlouf, A.S.H., Eds.; Springer: Cham, Switzerland, 2016; pp. 205–221.
5. Karapanagiotis, I.; Hosseini, M. Superhydrophobic coatings for the protection of natural stone. In *Advanced Materials for the Conservation of Stone*; Hosseini, M., Karapanagiotis, I., Eds.; Springer: Cham, Switzerland, 2018; pp. 1–25.
6. Manoudis, P.; Papadopoulou, S.; Karapanagiotis, I.; Tsakalof, A.; Zuburtikudis, I.; Panayiotou, C. Polymer-silica nanoparticles composite films as protective coatings for stone-based monuments. *J. Phys. Conf. Ser.* **2007**, *61*, 1361–1365. [CrossRef]
7. Manoudis, P.; Tsakalof, A.; Karapanagiotis, I.; Zuburtikudis, I.; Panayiotou, C. Fabrication of super-hydrophobic surfaces for enhanced stone protection. *Surf. Coat. Technol.* **2009**, *203*, 1322–1328. [CrossRef]
8. Manoudis, P.N.; Karapanagiotis, I.; Tsakalof, A.; Zuburtikudis, I.; Kolinkeová, B.; Panayiotou, C. Superhydrophobic films for the protection of outdoor cultural heritage assets. *Appl. Phys. A- Mater.* **2009**, *97*, 351–360. [CrossRef]
9. Chatzigrigoriou, A.; Manoudis, P.N.; Karapanagiotis, I. Fabrication of water repellent coatings using waterborne resins for the protection of the cultural heritage. *Macromol. Symp.* **2013**, *331–332*, 158–165. [CrossRef]
10. Facio, D.S.; Mosquera, M.J. Simple strategy for producing superhydrophobic nanocomposite coatings in situ on a building substrate. *ACS Appl. Mater. Interfaces* **2013**, *5*, 7517–7526. [CrossRef]
11. Aslanidou, D.; Karapanagiotis, I.; Panayiotou, C. Tuning the wetting properties of siloxane-nanoparticle coatings to induce superhydrophobicity and superoleophobicity for stone protection. *Mater. Des.* **2016**, *108*, 736–744. [CrossRef]
12. Helmi, F.M.; Hefni, Y.K. Using nanocomposites in the consolidation and protection of sandstone. *Int. J. Conserv. Sci.* **2016**, *7*, 29–40.
13. Pino, F.; Fermo, P.; La Russa, M.; Ruffolo, S.; Comite, V.; Baghdachi, J.; Pecchioni, E.; Fratini, F.; Cappelletti, G. Advanced mortar coatings for cultural heritage protection. Durability towards prolonged UV and outdoor exposure. *Environ. Sci. Pollut. Res.* **2017**, *24*, 12608–12617. [CrossRef] [PubMed]
14. Facio, D.S.; Carrascosa, L.A.M.; Mosquera, M.J. Producing lasting amphiphobic building surfaces with self-cleaning properties. *Nanotechnology* **2017**, *28*, 265601. [CrossRef]
15. Aslanidou, D.; Karapanagiotis, I.; Lampakis, D. Waterborne superhydrophobic and superoleophobic coatings for the protection of marble and sandstone. *Materials* **2018**, *11*, 585. [CrossRef] [PubMed]
16. Mosquera, M.J.; Carrascosa, L.A.M.; Badreldin, N. Producing superhydrophobic/oleophobic coatings on Cultural Heritage building materials. *Pure Appl. Chem.* **2018**, *90*, 551–561. [CrossRef]
17. Karapanagiotis, I.; Ntelia, E. Superhydrophobic Paraloid B72. *Prog. Org. Coat.* **2020**, *139*, 105224.

18. Tian, S.; Liu, S.; Gao, F.; Ren, J. Preparation and assessment of superhydrophobic organic-inorganic hybrid coatings for conservation of Yungang Grottoes. *Mater. Res. Soc. Symp. Proc.* **2011**, *1319*, 333–338. [CrossRef]
19. MacMullen, J.; Radulovic, J.; Zhang, Z.; Dhakal, H.N.; Daniels, L.; Elford, J.; Leost, M.A.; Bennett, N. Masonry remediation and protection by aqueous silane/siloxane macroemulsions incorporating colloidal titanium dioxide and zinc oxide nanoparticulates: Mechanisms, performance and benefits. *Constr. Build. Mater.* **2013**, *49*, 93–100. [CrossRef]
20. Cappelletti, G.; Fermo, P.; Camiloni, M. Smart hybrid coatings for natural stones conservation. *Prog. Org. Coat.* **2015**, *78*, 511–516. [CrossRef]
21. La Russa, M.F.; Rovella, N.; De Buergo, M.A.; Belfiore, C.M.; Pezzino, A.; Crisci, G.M.; Ruffolo, S.A. Nano-TiO$_2$ coatings for cultural heritage protection: The role of the binder on hydrophobic and self-cleaning efficacy. *Prog. Org. Coat.* **2016**, *91*, 1–8. [CrossRef]
22. Zarzuela, R.; Carbú, M.; Gil, M.L.A.; Cantoral, J.M.; Mosquera, M.J. Ormosils loaded with SiO$_2$ nanoparticles functionalized with Ag as multifunctional superhydrophobic/biocidal/consolidant treatments for buildings conservation. *Nanotechnology* **2019**, *30*, 345701. [CrossRef]
23. Taglieri, G.; Daniele, V.; Del Re, G.; Volpe, R. A new and original method to produce Ca(OH)$_2$ nanoparticles by using an anion exchange resin. *Adv. Nanoparticles* **2015**, *4*, 17–24. [CrossRef]
24. Galván-Ruiz, M.; Hernández, J.; Baños, L.; Noriega-Montes, J.; Rodríguez-García, M.E. Characterization of calcium carbonate, calcium oxide, and calcium hydroxide as starting point to the improvement of lime for their use in construction. *J. Mater. Civil Eng.* **2009**, *21*, 625–708. [CrossRef]
25. Gunasekaran, S.; Anbalagan, G.; Pandi, S. Raman and Infrared spectra of carbonates of calcite structure. *J. Raman Spectrosc.* **2006**, *37*, 892–899. [CrossRef]
26. Manoudis, P.N.; Karapanagiotis, I. Modification of the wettability of polymer surfaces using nanoparticles. *Prog. Org. Coat.* **2014**, *77*, 331–338. [CrossRef]
27. Karapanagiotis, I.; Manoudis, P.N.; Savva, A.; Panayiotou, C. Superhydrophobic polymer-particle composite films produced using various particle sizes. *Surf. Interface Anal.* **2012**, *44*, 870–875. [CrossRef]
28. Pedna, A.; Pinho, L.; Frediani, P.; Mosquera, M.J. Obtaining SiO$_2$–fluorinated PLA bionanocomposites with applicationas reversible and highly-hydrophobic coatings of buildings. *Prog. Org. Coat.* **2016**, *90*, 91–100. [CrossRef]
29. Gherardi, F.; Roveri, M.; Goidanich, S.; Toniolo, L. Photocatalytic nanocomposites for the protection of European architectural heritage. *Materials* **2018**, *11*, 65. [CrossRef]
30. Pargoletti, E.; Motta, L.; Comite, V.; Fermo, P.; Cappelletti, G. The hydrophobicity modulation of glass and marble materials by different Si-based coatings. *Prog. Org. Coat.* **2019**, *136*, 105260. [CrossRef]

© 2020 by the authors. Licensee MDPI, Basel, Switzerland. This article is an open access article distributed under the terms and conditions of the Creative Commons Attribution (CC BY) license (http://creativecommons.org/licenses/by/4.0/).

Article

Anti-Graffiti Behavior of Oleo/Hydrophobic Nano-Filled Coatings Applied on Natural Stone Materials

Mariateresa Lettieri [1], Maurizio Masieri [1], Mariachiara Pipoli [2], Alessandra Morelli [2] and Mariaenrica Frigione [2,*]

1. Istituto di Scienze del Patrimonio Culturale, CNR-ISPC, Prov.le Lecce-Monteroni, 73100 Lecce, Italy; mariateresa.lettieri@cnr.it (M.L.); maurizio.masieri@cnr.it (M.M.)
2. Department of Engineering for Innovation, University of Salento, 73100 Lecce, Italy; mchiara.pipoli@libero.it (M.P.); alessmorel@tiscali.it (A.M.)
* Correspondence: mariaenrica.frigione@unisalento.it; Tel.: +39-0832-297-215

Received: 7 October 2019; Accepted: 5 November 2019; Published: 7 November 2019

Abstract: In recent years, graffiti writings are increasingly regarded as a form of art. However, their presence on historic building remains a vandalism and different strategies have been developed to clean or, preferably, protect the surfaces. In this study, an experimental nano-filled coating, based on fluorine resin containing SiO_2 nano-particles, and two commercial products have been applied on compact and porous calcareous stones, representative of building materials used in the Mediterranean basin, and their anti-graffiti ability has been analyzed. All the tested experimental and commercial coatings exhibited high hydrophobicity and oleophobicity, thus meeting one of the basic requirements for anti-graffiti systems. The effects of staining by acrylic blu-colored spray paint and felt-tip marker were, then, assessed; the properties of the treated stone surfaces after cleaning by acetone were also investigated. Visual observations, contact angle measurements and color evaluations were performed to this aim. It was found that the protective coatings facilitated the spray paint removal; however high oleophobicity or paint repellence did not guarantee a complete cleaning. The stain from the felt-tip marker was confirmed to be extremely difficult to remove. The cleaning with a neat unconfined solvent promoted the movement of the applied polymers (and likely of the paint, as well) in the porous structure of the stone substrate.

Keywords: hydrophobic treatments; oleophobicity; nano-particles; stone protection; anti-graffiti coatings; chemical cleaning; acrylic-based paints; felt-tip markers

1. Introduction

Over the last few years, holistic approaches are trying to tackle the global graffiti phenomenon [1]. Any proposed solution is shared with all the involved stakeholders, including those who manage graffiti or utilize street art for city regeneration. The current studies are addressing not only to the graffiti removal, but also to the knowledge and protection of the street art murals [2–4].

However, graffiti on building façades, especially those with a cultural and historical value, still remain a vandalism [5]. Several strategies are used to either remove these graffiti or protect the surfaces against their harmful effects [6].

Mechanical, chemical or laser techniques are typically employed for the cleaning procedures [7–11]; biocleaning methods have been also proposed [12,13]. Graffiti removal is expensive and, in some cases, may cause stone damage due to chemical contamination, by-products formation and physical changes. Therefore, preventive actions are preferred, especially on artifacts of historical and artistic relevance.

To this aim, coatings acting as anti-graffiti barrier have been developed in the last years and many products are nowadays commercially available. The anti-graffiti products can be grouped into three main classes: sacrificial, semi-permanent and permanent [14]. The sacrificial coatings are removed during the cleaning process and they need to be renewed; the semi-permanent systems endure a few cleaning cycles (normally, not more than two or three); the permanent products are not taken away during the cleaning process and they are able to withstand several cleaning cycles. The formulations suitable for application on stone materials [15–23] are mainly based on waxes, fluorinated polymers, silicon resins or polyurethanes; more recently, coatings incorporating nano-particles [17,24–26], organic–inorganic hybrid products [27–29] and surface functionalization [30] have been investigated as potential anti-graffiti systems. The fundamental characteristics that these treatments must display are: transparency, permeability to water vapor, durability under outdoor conditions. In addition, non-wettable coatings are preferred to enable the treated surfaces to repel paints and other staining agents [31].

Wetting is the ability of a liquid to maintain contact with a solid surface; the balance between the intermolecular interactions of adhesive type (liquid to surface) and cohesive type (liquid to liquid) controls this feature. Wettability of a solid surface can be described by the contact angles and sliding angle; models (Young, Wenzel, Cassie–Baxter) have been developed to illustrate the wetting phenomena on surfaces [32,33]. The decrease of the wettability corresponds to an increase in liquid-phobicity. Regarding water, a surface is called hydrophobic when water drops on it exhibit contact angles greater than 90°. When the contact angles are 150° or higher, the surface is classified as superhydrophobic. Despite their high level properties, superhydrophobic surfaces have low mechanical wear resistance and poor long-term durability, therefore, their utilization in real applications is still limited [33–35]. Water-repellency is achieved in super-hydrophobic surfaces with water contact angle hysteresis <10°, that is a low difference between the advancing and receding contact angles. In these conditions, a water droplet can move with little applied force and easily rolls off from the surface [36]. Similar considerations are applied to oleo-phobic/-repellent surfaces [37].

Several methods have been used to design and produce non-wettable stone surfaces, often bioinspired to either plants or insects surfaces. The most common procedures include the application of polymer coatings able to reduce the surface tension of the substrate, sol-gel processes and controlled nano-particle embedding into polymer matrices [29].

Hydrophobicity and oleophobicity are assumed as basic properties to provide protection against graffiti [38–46]; however, the related actual anti-graffiti action is taken for granted and few are the studies where this property has been proved in oleo/hydrophobic coatings applied on stone substrates [43,45]. Starting from these issues, an experimental work, aimed at investigating the performance of oleo/hydrophobic coatings and their behavior as anti-graffiti systems applied on building stone materials, has been undertaken. The present study is a part of a wide research on products suitable for superficial protection of stone materials.

Three products have been applied on both compact and porous calcareous stones representative of building materials used in the Mediterranean basin. Two of the used products are already commercially available and are suggested to provide water and anti-graffiti protection to stone surfaces. An experimental nano-filled system, based on fluorine resin containing SiO_2 nano-particles, has been also investigated.

Blu-colored spray paint and felt-tip marker were used as staining agents. Traditional and easy-to-apply methods usually used by restores and professionals were chosen for the stain removal. Cleaning procedures by warm water and, subsequently, by acetone, were applied to the stained stone surfaces. The properties of the stone surfaces were, then, analyzed by visual observations, contact angle measurements and color evaluations, on stained and neat surfaces, as well as after the paint removal. The percentage of residual stain and the efficacy of the cleaning procedures, in comparison with the unprotected stone surfaces, were used to measure the paint removal.

2. Materials and Methods

2.1. Protective Products, Stone Materials, and Staining Agents

An experimental formulation and, for comparison purposes, two commercial products were investigated.

The experimental product, hereinafter nanoF, is a water-based fluorine resin (12.7 wt.%) containing SiO_2 nano-particles (10 wt.%), 40–50 nm in dimensions (supplied by Kimia S.p.A., Perugia, Italy); nanoF has density of 1.04 g/cm^3 and pH between 7 and 8; the viscosity, similar to those of the commercial systems, was appropriate for the application by brush [47]. The first commercial product, F (trade mark Fluoline PE, supplied by C.T.S. S.r.l., Altavilla Vicentina, Italy), is an aqueous dispersion of fluoropolyethers (10 wt.%); this system was selected to compare the experimental nano-filled formulation to a unfilled chemically similar product already on the market (nanoF and F are, indeed, both fluorine-based). The second commercial system, hereinafter SW (trade mark Kimistone DEFENDER, supplied by Kimia S.p.A., Italy), consisting of a mixture of organic silicon compounds and microcrystalline waxes in water solution; SW was included in the study because it belongs to a family of products, i.e., the silicon-based, widely and successfully used in the field of stone conservation. According to the technical sheets, both the commercial systems are able to provide a reversible and hydrophobic coating on the treated surfaces, with dirt-repellent and anti-graffiti properties. Additional information about the protective systems has been reported in a previous study [47].

The three protective products were tested on two natural calcareous stone materials, representative of construction materials used for historic and civil buildings in many countries in the Mediterranean basin. A highly porous calcarenite (PS), named "Lecce stone", and a compact limestone (CS), known as "Trani stone", were used.

The principal constituent of "Lecce stone" is calcite (93%–97% [48]); in this material, very small quantities of clay, phosphates and other non-carbonate minerals are also present [49,50]. Petrographically, "Lecce stone" is a packstone [51], made of microfossils and fossil remains within a groundmass of fine calcareous detritus (Figure 1a). The used samples exhibited a porosity of 39%, with pore sizes mainly between 0.5 and 6 µm (Figure 1b), as analyzed by mercury-intrusion porosimetry (MIP) [52].

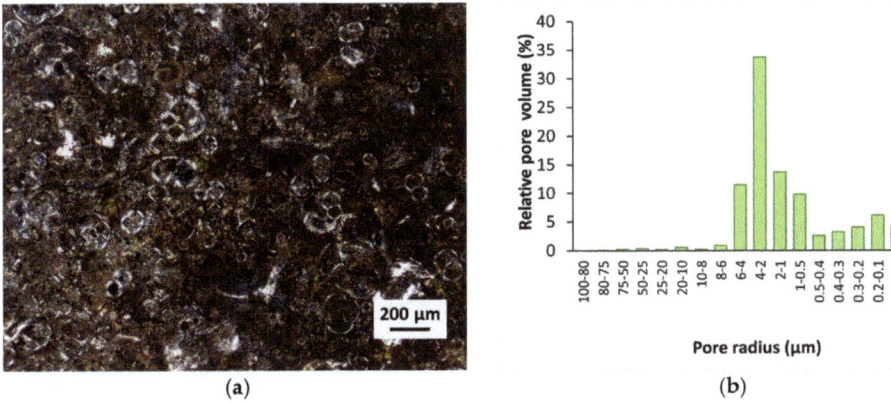

Figure 1. Lecce stone: (**a**) image taken on thin sections through optical microscope (Eclipse LV 100 PL, Nikon, Tokyo, Japan) in transmitted light; (**b**) pore-size distribution curve.

"Trani stone" is mainly composed of calcite (>95% [36]), and low amounts of clay minerals and iron oxides [33]. "Trani stone" is a packstone [51] made of calcareous detritic grains, very well cemented by a crystalline cement filling the interparticle porosity (Figure 2a). The open porosity, measured by MIP, was very low (2%); the pore size was mainly between 0.025 and 0.001 µm (Figure 2b).

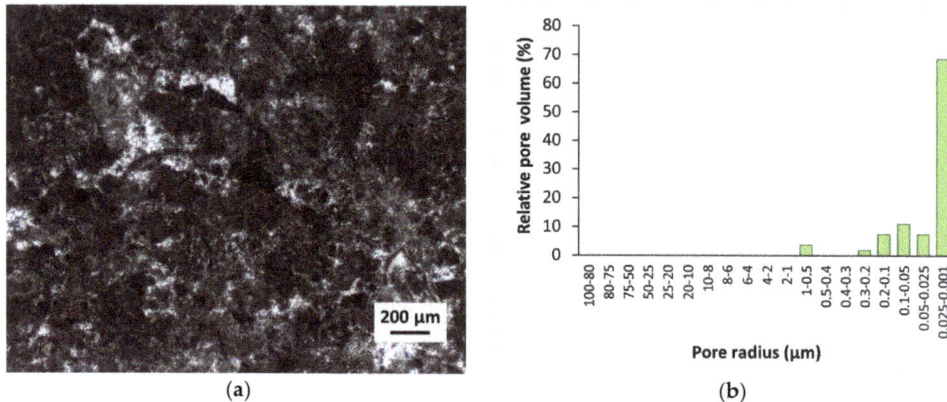

(a) (b)

Figure 2. Trani stone: (**a**) image taken on thin sections through optical microscope (Eclipse LV 100 PL, Nikon, Tokyo, Japan) in transmitted light; (**b**) pore-size distribution curve.

Two staining agents were applied to test the protective action of each coating applied on the stones, i.e., (1) a commercial acrylic spray paint (Cilvani RAL by Cilvani S.r.l., Caivano, Italy), blu-colored (RAL code 5015), provided in a pressurized can; (2) a water-based acrylic paint marker (POSCA by UNI Mitsubishi pencil, Tokyo, Japan), blu-colored (RAL code 5005), having a bullet tip 1.8–2.5 mm wide (PC-5M).

2.2. Stone Specimens

Prismatic specimens of PS and CS stones, with dimensions of 5 cm × 5 cm × 1 cm, were cut by a saw from quarry blocks. According to the UNI10921 standard protocol [53], the samples were smoothed with abrasive paper (180-grit silicon carbide), cleaned with a soft brush and washed with deionized water in order to remove dust deposits. The stone specimens were completely dried in oven at 60 °C, until the dry weight was achieved, and stored in a desiccator with silica gel (relative humidity (R.H.) = 15%) at 23 ± 2 °C. Before the application of each protective product, the stone specimens were conditioned in equilibrium with the surrounding environment (24 h in the laboratory, at 23 ± 2 °C and 45% ± 5% R.H.).

The treatments were applied by brush on 3 sample surfaces (5 cm × 5 cm) for each product.

For F and SW, the amounts of product suggested in the technical sheets were applied. Preliminary tests were used to verify the optimal amount of nanoF to effectively treat the two stone materials [47]. Following the minimum intervention criteria, the quantities suitable to obtain highly hydrophobic surfaces along with minimal color changes were identified. For all the products, greater amounts were necessary to guarantee good performances in the highly porous stone (PS).

The actual amount of the applied product was evaluated by weighting the specimens before and after the treatment. After the application of the products, all the specimens were kept in the laboratory at 23 ± 2 °C and 45% ± 5% R.H. for 30 days; then, they were dried in oven at 40 °C until the weight stabilization was achieved, the stabilization being controlled by periodical weight measurements. The treatments' harmlessness, assessed in terms of surface color variations and reduction in water vapor permeability, was proved in a previous study [47]; the main results are summarized in Table 1.

Table 1. Amount of applied product, color change (ΔE^*_{ab}) and variation of water vapor permeability (ΔP) evaluated after the protective treatment [47].

Stone Substrate	Product	Applied Amount (g/m^2)	ΔE^*_{ab} (CIELAB unit)	ΔP (%)
	nanoF	58	1.72	+15
CS	F	60	2.50	−5
	SW	109	2.56	−38
	nanoF	155	1.38	+14
PS	F	160	3.97	−6
	SW	313	3.51	−51

The staining of the surfaces was performed 2 months after the application of the protective coatings, as detailed in Section 2.3.

During the preparation of the specimens, their subsequent treatments and relative tests, the environmental conditions were monitored by means of a thermo-hygrometer (Mod. EMR812HGN, Oregon Scientific, Hong Kong, China). This instrument is able to collect temperature data from −50 to 70 °C (with resolution of 0.1 °C) and relative humidity data in the range 2%–98% (with resolution of ±1%). All weight measurements were registered using an analytical balance (Model BP 2215, Sartorius, Goettingen, Germany) with an accuracy of ±0.1 mg.

All the procedures carried out on the stone samples are illustrated in Figure 3.

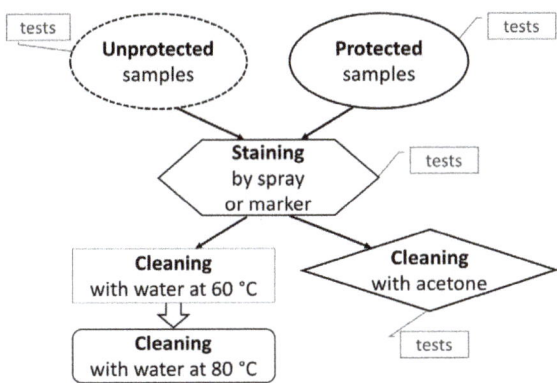

Figure 3. Experimental procedures.

2.3. Staining Methods and Removal Procedures

2.3.1. Spray Paint

The staining with the spray paint was carried out on untreated and protected stone samples, conditioned in equilibrium with the surrounding environment (24 h in the laboratory, at 23 ± 2 °C and 50% ± 5% R.H.). Two coats of paint were sprayed on specimens placed on a 45° tilted surface (Figure 4a). The distance between the sample surface and the nebulizer was about 15 cm. In order to limit the deposition of paint to an area of 1.5 cm × 5 cm, the staining was performed with the aid of a stencil and the lateral sides of the specimens were protected with a polyester (PET) film.

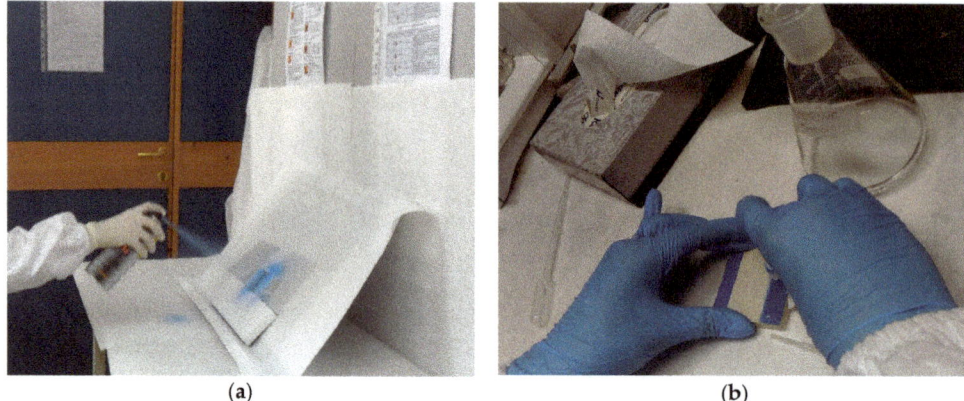

Figure 4. Staining by spray paint (**a**) and removal with acetone (**b**).

After the application of the paint, the samples were stored for 2 days in the laboratory at 23 ± 2 °C and 50% ± 5% R.H. The removal procedures were performed 20 days after the staining. Starting from the data reported in the technical sheet of the SW product, a first cleaning with warm water (at 60 °C and next at 80 °C) and paper towels, was firstly attempted on testing areas. Since this method resulted totally ineffective, it was no more applied to the stained specimens and its effects were not further investigated. Then, following the recommendation reported in the international code [54], cleaning with an organic solvent was tried. Acetone was selected because it is (alone or in mixture with other organic solvents) a traditional solvent for the cleaning of stone materials affected by graffiti [6]; it has been successfully used to remove acrylic paints, as reported in previous studies [9,55]. Acetone analytical grade, supplied by Carlo Erba Reagents (Val de Reuil, France) was used. For each sample, a wet paper towel was rubbed across the stained area (Figure 4b) for 25 complete back and forth cycles [54]; the towel was dunked in acetone every 5 cycles.

2.3.2. Felt-Tip Paint Marker

The staining with the felt-tip marker was performed on untreated and protected stone samples conditioned in equilibrium with the surrounding environment (24 h in the laboratory, at 23 ± 2 °C and 50% ± 5% R.H.). The paint was applied to an area of approximately 1.5 cm × 5 cm in the same specimens stained with the spray paint, but in a different zone (Figure 5).

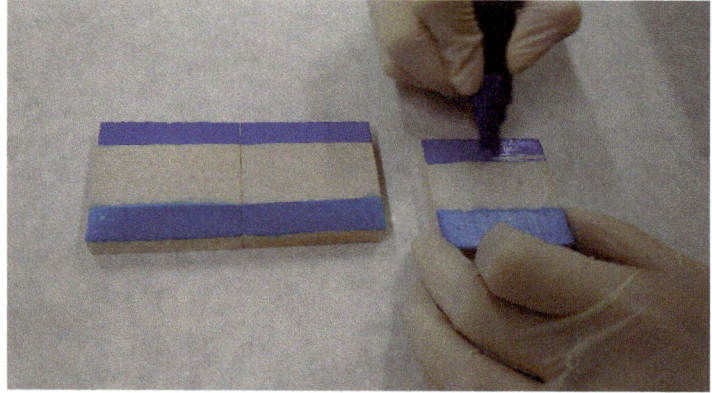

Figure 5. Staining by felt-tip marker.

The samples were, then, kept in the laboratory conditions (23 ± 2 °C and 50% ± 5% R.H) for 2 days. The same removal procedures used to clean the samples stained with the spray paint (described in Section 2.3.1) were carried out 18 days after the staining with the marker.

2.4. Analytical Investigations

In order to evaluate hydrophobicity and oleophobicity of the stone surfaces, static contact angle measurements were performed before and after the coatings' application. A Costech apparatus was used to deposit micro-drops of the wetting liquid on the stone surfaces. The shape of the drop was recorded with a camera and the related contact angle was calculated by means of the "anglometer 2.0" software (Costech). To assure the reproducibility of the test, the image of each drop was acquired 15s after its deposition.

The water-stone contact angles were measured on 30 different positions of the surface of each specimen using deionized water as wetting liquid, according to the European standard [56]. A commercial olive oil (purchased from a local market) was used to determine the oil-stone static contact angle, following a procedure already proposed in other studies [57–59]; 5 measurements were performed on each sample and the results were averaged.

For the unprotected Lecce stone (PS), the absorption of wetting liquids is rapid because of the high stone porosity. Consequently, during the test, the drops of both water and oil were suddenly absorbed inside the stone and the contact angle was not determinable.

The water-stone contact angle measurements were repeated after the staining and after the paint removal.

Color measurements [60] were performed with a spectrophotometer (mod. CM-700d, Konica Minolta Sensing, Singapore), using CIE Standard illuminant D65 and the target mask 8 mm in diameter. Ten measurements were performed on each sample area and the instrument was recalibrated to a white calibration cap at the start of each measurement session. The color coordinates were measured on the unprotected surfaces, after the coatings' application, after the staining and after the paint removal.

The colour changes (ΔE^*_{ab}) were calculated through the $L^*a^*b^*$ (CIE 1976) system, using Equation (1):

$$\Delta E^*_{ab} = [(\Delta L^*)^2 + (\Delta a^*)^2 + (\Delta b^*)^2]^{1/2} \tag{1}$$

where L^* is the lightness/darkness coordinate, a^* the red/green coordinate ($+a^*$ indicating red and $-a^*$ green), and b^* the yellow/blue coordinate ($+b^*$ indicating yellow and $-b^*$ blue).

All the colour variations were determined by the comparison with the untreated surfaces, using the averaged values of L^*, a^*, and b^* for each sample.

The residual stain (RS) after cleaning was evaluated as a percentage by Equation (2):

$$RS = [(\Delta E^*_{ab})c/(\Delta E^*_{ab})s] \times 100\% \tag{2}$$

where $(\Delta E^*_{ab})c$ is the colour variation of the cleaned surfaces and $(\Delta E^*_{ab})s$ is the colour variation of the stained surfaces.

In addition, the efficacy of the cleaning procedure, as compared with the unprotected surfaces, was evaluated by Equation (3):

$$\text{Relative Efficacy\%} = [(RS_u - RS_t)/RS_u] \times 100\% \tag{3}$$

where: RS_u and RS_t are the residual stain values, as defined in Equation (2), for the unprotected and protected surfaces, respectively.

The stained surfaces were examined under a binocular stereomicroscope (Stemi SV11, Zeiss, Oberkochen, Germany) at magnifications of up to 100×.

3. Results and Discussion

3.1. Assessment of Basic Prerequisite for Antigraffiti Protection

The products used as an anti-graffiti barrier typically have a low surface energy at interface, which results in minimizing the contact with the applied paints or inks mainly because the protected surfaces become water and oil repellent. Therefore, treatments able to give good hydrophobicity and oleophobicity to the surfaces are assumed to act as effective systems of graffiti protection [45].

In our case, after the coating's application, most of the protected stone surfaces were able to repel both water and oil, even if to different extent, as it can be clearly inferred from the observation of images reported in Figure 6. Oleophobicity was not observed only in the case of the SW coating, irrespective of the stone substrate.

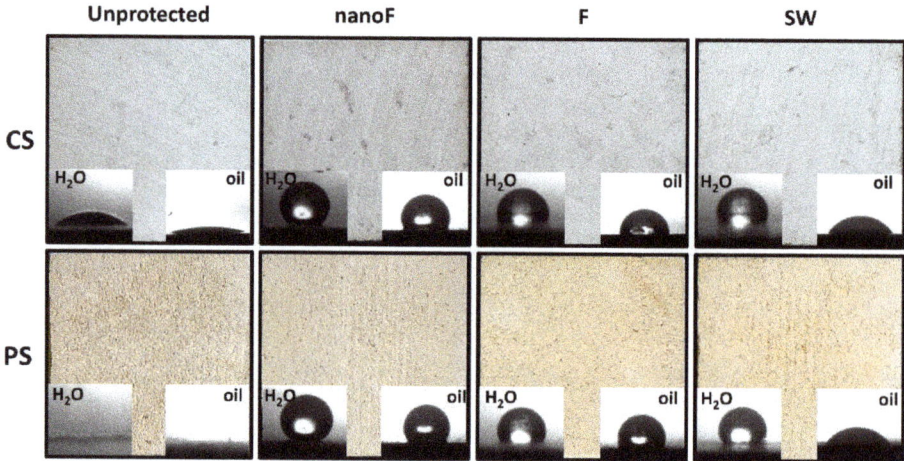

Figure 6. Water and oil droplets on the stone surfaces before and after the protective treatments.

The contact angle values (reported in Table 2) confirmed these surface properties, since water-stone contact angles greater than 90° are typical of hydrophobic surfaces, while oil-stone contact angles above 70°–80° account for oleophobicity. The lower values measured for oil contact angles with respect to water ones are due to the low surface tension of the oil drops (32 mN/m for olive oil [61], 72 mN/m for water [57]).

Table 2. Water-stone static contact angle (WCA) and oil-stone static contact angle (OCA), both in degrees, measured on unprotected and protected PS and CS stone surfaces, with the indication of standard deviation.

Samples	CS		PS	
	WCA	OCA	WCA	OCA
Unprotected	40 ± 8	13 ± 1	Not determinable	Not determinable
nanoF	139 ± 5	114 ± 1	142 ± 5	122 ± 7
F	106 ± 4	93 ± 4	119 ± 3	114 ± 4
SW	114 ± 4	56 ± 1	122 ± 4	56 ± 2

3.2. Staining by Spray Paint and Removal

The superficial distribution of the spray paint on the stone surfaces was observed through the stereomicroscope. The images recorded by the instrument on the different specimens are illustrated in Figure 7.

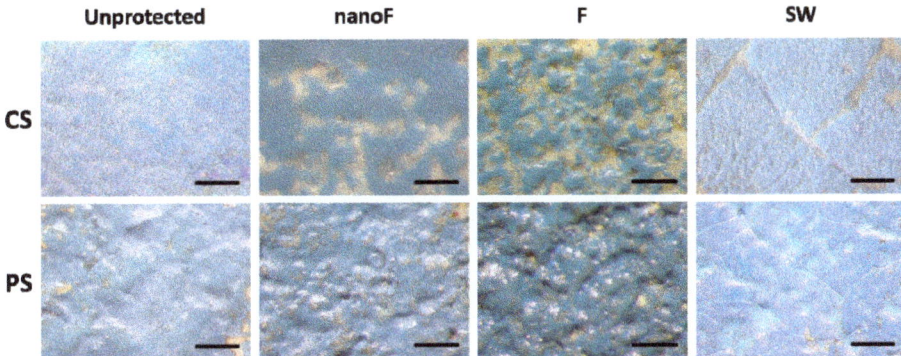

Figure 7. Images taken by the stereomicroscope of the surfaces stained by spray paint: comparison between the protected and unprotected surfaces (CS = compact stone; PS = porous stone). Each scale bar indicates 200 µm.

In the case of the untreated CS specimens, the paint covered the surface with a uniform film, totally hiding the stone beneath. The presence of protective coatings modified the distribution of the paint. In particular, on the samples treated with SW, the paint coating was affected by cracks due to the drying shrinkage. This latter was likely restrained by the protective layer since cracking was not observed for the unprotected samples. Passing to analyze the specimens treated with nanoF and F products, the protected surfaces seemed to repel the spray paint, which mainly arranged as separate droplets [15,25,40,62–64], leaving several portions of the stone uncovered. The distribution of the paint affected the color variations, leading to lower differences where the staining agent was repelled, as can be deduced from the data reported in Table 3.

Table 3. Global color difference (ΔE^*_{ab}) determined after the staining by spray paint and after the cleaning. For each data set the standard deviation is reported.

Samples	CS		PS	
	After Staining	After Removal	After Staining	After Removal
Unprotected	54.53 ± 0.33	20.19 ± 2.27	55.39 ± 1.14	44.58 ± 4.36
nanoF	40.72 ± 3.56	21.72 ± 1.85	55.73 ± 3.96	37.11 ± 3.35
F	45.23 ± 5.18	5.99 ± 1.58	56.13 ± 2.61	13.78 ± 1.53
SW	55.28 ± 0.44	15.27 ± 1.95	59.51 ± 1.99	19.93 ± 2.96

The different appearance is also in agreement with the surface oleophobicity. In fact, oil-stone contact angles higher than 90° were measured only on the nanoF and F treated samples (114° and 93°, respectively).

Comparable behaviors were observed in the case of PS specimens. A uniform coverage of the stone surface was observed for the unprotected specimens, while a paint coating with cracks was found on the samples coated by SW product. Although less noticeable, uncovered portions of stone and repellence against the paint were observed on the specimens protected by nanoF and F systems, which exhibited the highest oleophobicity. Actually, the high porosity along with the large pore radius likely promoted the accumulation of paint inside the cavities of the porous stone, making less evident the paint repellence. In addition, these effects equalized the color differences and, thus, comparable ΔE^*_{ab} were measured in the stained PS samples (see results reported in Table 3).

The cleaning procedure of this kind of stain with acetone did not give good results. The paint removal was unsuccessful, as already clear to the visual inspection by the naked eye (Figure 8).

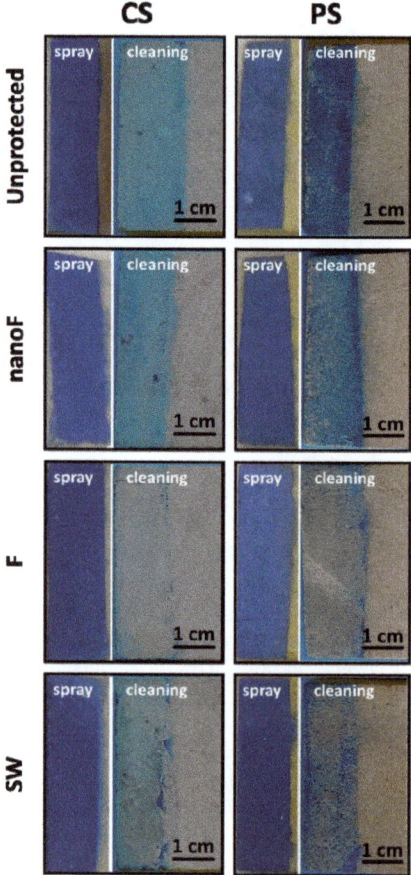

Figure 8. Protected and unprotected stone samples after the staining by spray paint and after the cleaning procedure (CS = compact stone; PS = porous stone); the neat surface is on the right of each image.

To this regard, different methods have been proposed to assess the efficacy of stain removal [26,65,66]. A first one was based on the ΔE^*_{ab} measured after the cleaning: values below 5 account for adequate cleaning, while values higher than 10 cannot be accepted; for ΔE^*_{ab} between 5 and 10, the color variations are well visible but still tolerable. In addition, RS can be used to evaluate the removal efficacy [67,68]: values below 10% can be judged suitable; RS above 20% means ineffective stain removal; RS values between 10% and 20% are not optimal, but tolerable.

Taking into account these classifications, the cleaning with acetone resulted acceptable only for the CS samples treated with the F product, where ΔE^*_{ab} of approximately 6 CIELAB units and RS of 13% were measured. Nevertheless, the application of a protective layer was helpful in facilitating the removal of the spray paint. As illustrated in Figure 9a, the RS percentages were always lower than those calculated for the unprotected specimens, except for CS treated with nanoF for which comparable values were found.

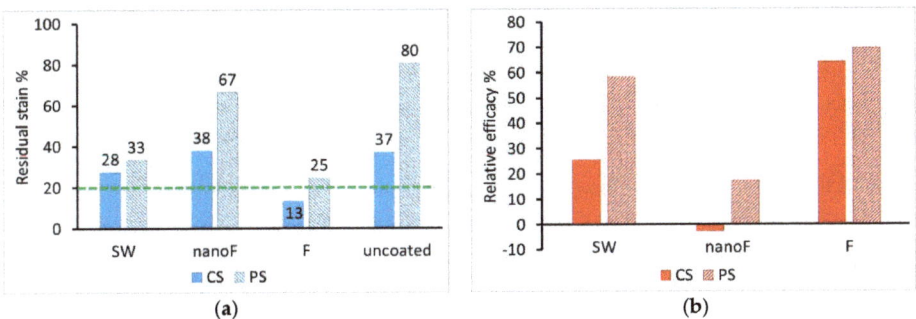

Figure 9. Evaluation of the spray paint removal: (**a**) residual stain, as defined in Equation (2), with the green line indicating the acceptable threshold; (**b**) relative efficacy, as defined in Equation (3).

The relative efficacy, reported in Figure 9b, of the complete procedures (i.e., protection + cleaning) was found to be good in the case of the F-treated specimens; on the other side, it resulted low in the case of the samples where the nanoF coating was applied. Overall, higher values were found for the porous stone, where the paint removal was appreciably more difficult without the presence of a protective layer.

The results of the cleaning seemed to be not related to the observed superficial morphology of the sprayed paint. Neither high oleophobicity nor paint repellence assured good outcomes. On the other hand, as also found in other studies, an observed repellence against the paint did not assure the complete stain removal [69].

The contact angle measurements, whose results are reported in Table 4, were able to supply important information.

Table 4. Water-stone contact angles (in degrees) before and after the protective treatments, after the staining by spray paint, and after the cleaning. For each data set the standard deviation is reported.

Stone Support	Samples	Before Treatments	After Protection	After Staining	After Removal
CS	Unprotected	55 ± 10	–	93 ± 1	76 ± 3
	nanoF	38 ± 7	140 ± 5	98 ± 3	115 ± 4
	F	44 ± 7	105 ± 5	92 ± 1	116 ± 4
	SW	39 ± 7	112 ± 3	101 ± 4	111 ± 4
PS	Unprotected	n.d.	–	98 ± 5	43 ± 14
	nanoF	n.d.	144 ± 4	106 ± 7	126 ± 4
	F	n.d.	120 ± 3	95 ± 3	128 ± 4
	SW	n.d.	121 ± 4	101 ± 4	120 ± 4

n.d. = not determinable.

The staining caused a reduction in contact angles in all the samples, with values comparable to those measured on the unprotected surfaces and consistent with the presence of a superficial layer of paint. After the removal with acetone, the contact angle values were still high and measurable. It can be concluded that neither the stain nor the protective coatings were removed, since the hydrophobic character of the surfaces was modified only to a limited extent. An increase in the contact angle values was even observed in the case of the F-treated samples after the cleaning. Other studies report similar behaviors [45,65,70], but the related mechanisms are not clearly explained. A possible explanation for this behavior is the so-called "reverse migration" [71–73]: the used solvent penetrates into the pores of the stone, but migration towards the external surface occurs for evaporation. During these processes, the solvent is able to carry the polymer protective with it. Consequently, greater amounts of product can reach the surface, accounting for a higher hydrophobicity. Even if in fine-porous stone materials (as

CS) this phenomenon should be restricted [74,75], comparable behaviors were observed irrespective of the porosimetric features of the stone samples. On the other hand, the type of solvent, its tendency to dissolve polymers, and the drying conditions mainly affect the "reverse migration" [75–77]; in addition, the dimensions of the molecules and the solution viscosity can influence the movements of the polymer inside the stone structure.

3.3. Staining by Felt-Tip Marker and Removal

The surface appearance of the stone stained by felt-tip marker was investigated through the stereomicroscope, that recorded the images reported in Figure 10.

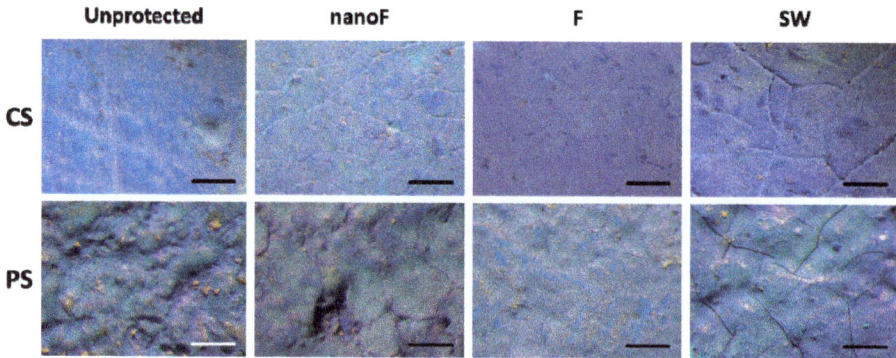

Figure 10. Images taken by the stereomicroscope of the surfaces stained by felt-tip marker: comparison between the protected and unprotected surfaces (CS = compact stone; PS = porous stone). Each scale bar indicates 200 μm.

After the staining, the paint applied by felt-tip marker considerably covered the stone surfaces. In the unprotected samples, the morphological features of the stone were still recognizable and accumulation of the paint in the cavities of the surface was limited. Conversely, in the protected samples, the staining by the marker produced films affected by cracks, as already found in other studies [25]. Additionally, accumulation of stain within the superficial hollows was observed in the PS protected samples.

Dissimilar craquelure patterns were observed on the different specimens. More fractured films were seen in the case of the nanoF and SW treated-samples, regardless of the stone substrate. The paint applied on the F-treated samples displayed sporadic and less evident cracks, instead. Generally speaking, the thickness of any kind of coating influences morphologies and fragment size of cracks [78,79], with a critical value below which no fracturing occurs [79,80]. The limited cracking on the F treated-samples suggested a lower thickness of the paint level, probably due to absorption by the protective coating.

The stain applied by the felt-tip marker was not removed with acetone from the stone surfaces; neither the protective coatings improved the cleaning results. In fact, markers are considered the most aggressive staining agents among the methods for graffiti writings. This is mainly due to the fact that their inks contain high percentages of solvents, are very fluid and are, thus, prone to easily fill the pores of the substrate [81,82] already during the staining action.

Following the evaluation detailed in Section 3.2, the stain removal with acetone cannot be judged effective, as also visible to the naked-eye observation (Figure 11).

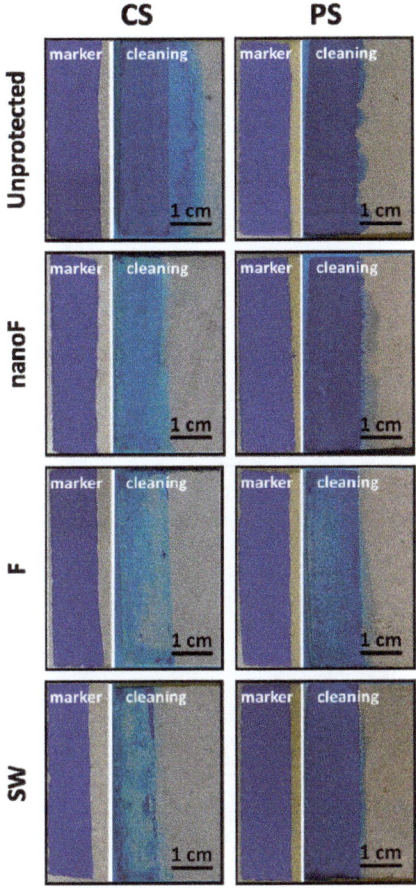

Figure 11. Protected and unprotected stone samples after the staining by felt-tip marker and after the cleaning procedure (CS = compact stone; PS = porous stone); the neat surface is on the right of each image.

Accordingly, unacceptable ΔE^*_{ab} were measured after the cleaning of all the samples (see data reported in Table 5), with RS values (Figure 12a) much greater than the tolerable threshold (i.e., 20%). Only the treatments with the F product yielded low RS percentages.

Table 5. Global color difference (ΔE^*_{ab}) determined after the staining by felt-tip marker and after the cleaning. For each data set the standard deviation is reported.

Samples	CS		PS	
	After Staining	After Removal	After Staining	After Removal
Unprotected	66.19 ± 0.37	50.65 ± 1.98	71.22 ± 0.70	65.03 ± 1.93
nanoF	58.52 ± 0.46	23.67 ± 9.57	70.61 ± 0.56	66.33 ± 0.52
F	57.95 ± 1.83	13.35 ± 4.40	72.88 ± 0.32	41.23 ± 12.22
SW	48.24 ± 1.87	19.81 ± 6.49	72.38 ± 0.57	62.60 ± 2.99

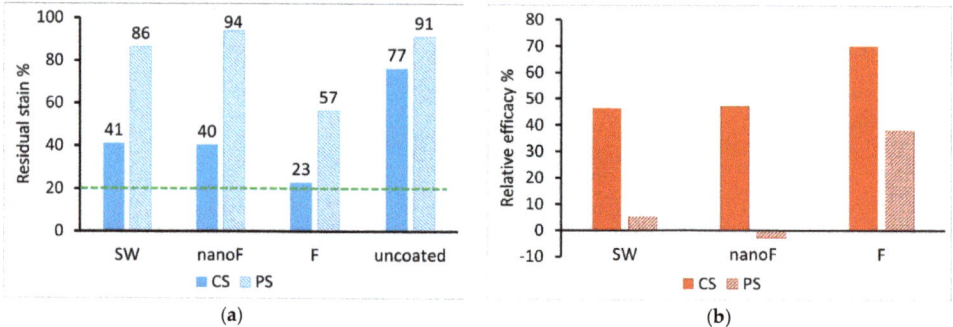

Figure 12. Evaluation of the felt-tip marker removal: (**a**) residual stain, as defined in Equation (2), with the green line indicating the acceptable threshold; (**b**) relative efficacy, as defined in Equation (3).

This result might arise from the embedding of the paint into the protective coating, as previously described. The incorporation of the stain enhanced, even if only slightly, the anti-graffiti effectiveness [68,83]. The relative efficacy showed that the cleaning procedure worked better on the CS samples, unlike the spray paint removal, as witnessed by the results presented in Figure 12b.

The correlation of these results with those relative to the oleophobicity of the surfaces did not show a clear trend. Anyway, the most oleophobic surfaces (i.e., that treated with nanoF) were not cleaned to a larger extent.

The results of the contact angle measurements, reported in Table 6 confirmed the failed removal of both paint and protective coating.

Table 6. Water-stone contact angles (in degrees) before and after the protective treatments, after the staining by felt-tip marker, and after the cleaning. For each data set the standard deviation is reported.

Stone Support	Samples	Before Treatments	After Protection	After Staining	After Removal
CS	Unprotected	55 ± 10	–	89 ± 2	92 ± 3
	nanoF	38 ± 7	140 ± 5	80 ± 3	101 ± 9
	F	44 ± 7	105 ± 5	79 ± 5	106 ± 2
	SW	39 ± 7	112 ± 3	80 ± 3	110 ± 2
PS	Unprotected	n.d.	–	98 ± 5	106 ± 7
	nanoF	n.d.	144 ± 4	83 ±13	122 ± 8
	F	n.d.	120 ± 3	90 ± 12	117 ± 4
	SW	n.d.	121 ± 4	70 ± 12	106 ± 6

n.d. = not determinable.

After the staining, contact angles between 70° and 97° were measured. Hydrophobicity was recovered on the cleaned samples, but the contact angles remained below the values obtained after the application of the protective treatment. The phenomenon of "reverse migration" can be hypothesized also in these samples. This effect, particularly evident for the unprotected surfaces, where unexpected high contact angles (>90°) were found after the cleaning, was most likely due to transfer of the absorbed paint binder activated by the cleaning solvent towards the surface on the stone. On the other hand, the occurrence of movements of the solvent and the dissolved stain were well visible to the naked-eye, since, during the cleaning, the paint spread beyond the previously stained area (Figures 9 and 11).

4. Conclusions

The anti-graffiti behavior of three protective treatments has been tested on compact and porous calcareous stone materials. An experimental formulation, based on fluorine resins and

SiO_2 nano-particles, and two commercial products were applied. The results of cleaning procedures after the staining by spray paint and felt-tip marker were discussed.

The applied products provided surfaces able to repel both water and oil, thus meeting a fundamental requirement for anti-graffiti systems.

The simulation of staining actions gave rise to a distribution of the paints on the surfaces dependent on the presence and nature of the protective coating. Uniform films covering the surface, totally hiding the stone beneath, were observed in the case of untreated specimens. In most of the protected surfaces, the paint film was affected by cracks due to its shrinkage upon drying. The protective coatings restrained this cracking, which, in fact, was not noticed for the unprotected samples. Evidence of repellence against the stain was optically observed only after the application of the spray paint. In accordance with the higher surface oleophobicity, separate droplets of paint, together with portions of uncovered stone, were seen on the nanoF and F treated-samples. The morphological and porosimetric features of the stone seemed influence the paint spreading on the surface to a limited extent.

Although the removal with acetone affected the paint layer to different degrees, this procedure was not able to supply good results. The residual stain percentage was always above the threshold limit of 20%, except for the compact stone surfaces coated with F product and stained by spray paint. The paint removal was more difficult on the highly porous stone, where greater values of residual stain were found. The presence of protective coatings enhanced the spray paint removal On the contrary, the stain from the felt-tip marker was not removed even where protective coatings were applied, thus confirming a stronger action of this graffiti agent. The relative efficacy of the cleaning was influenced by the type of staining agent rather than the porosity of the substrate. In both cases, the proved hydrophobicity and oleophobicity, as well as the observed paint repellence, did not provide positive outcome of the cleaning.

The protective coatings were not eliminated from the surfaces during the cleaning; rather, the polymer protectives migrated into the porous structure of the stone under the effect of solvent evaporation. In fact, as a consequence of greater amounts of hydrophobic product moved at the surface, increased contact angle values were measured after the cleaning with acetone. This result further supports the inefficacy of the cleaning procedure with neat acetone. The unconfined solvent can spread the dissolved paint into the pores of the substrate. In this case, the stain may affect also the internal part of the stone materials making almost impossible an effective cleaning.

In conclusion, although hydrophobicity and oleophobicity are considered basic requirements of anti-graffiti coatings, these surface properties do not assure good removal of vandalic writings. The success of the anti-graffiti action depends on the applied staining agent, on the used cleaning procedure and, to a more limited extent, on the affected substrate. Therefore, the actual performance of an anti-graffiti system cannot be deduced from the coating's properties but the effectiveness needs to be assessed in the specific applicative conditions.

Further investigations are in progress to verify the anti-graffiti efficacy under additional cleaning methods, specifically, chemical cleaners in gel matrices; the coatings' performance after the application of higher amounts of product is another ongoing evaluation.

Author Contributions: Conceptualization, M.L. and M.F.; Formal Analysis, M.L., M.M. and M.F.; Investigation, M.M., M.P. and A.M.; Writing—Review and Editing, M.L. and M.F.; Supervision, M.F.

Funding: This research received no external funding.

Acknowledgments: The work is the outcome of a scientific collaboration carried out with Kimia Company (Perugia, Italy), which supplied the experimental (nanoF) product under investigation. To this regards, the Authors wish to thank Assorestauro Association, and the Advisor of the Board Arch. Cristina Caiulo, for the scientific networking created between University of Salento and Kimia Company. Thanks are also due to Davide Melica for the realization of the thin cross sections.

Conflicts of Interest: The authors declare no conflict of interest.

References

1. GRAFFOLUTION, Awareness and Prevention Solutions against Graffiti Vandalism in Public Areas and Transport. SSP (Policy Oriented Research) of the Seventh European Programme of the European Commission. FP7-SEC-2013-1. 2016. Available online: http://project.graffolution.eu/po/ (accessed on 26 September 2019).
2. Macchia, A.; Ruffolo, S.A.; Rivaroli, L.; Malagodi, M.; Licchelli, M.; Rovella, N.; Randazzo, L.; La Russa, M.F. Comparative study of protective coatings for the conservation of Urban Art. *J. Cult. Herit.* **2019**, in press. [CrossRef]
3. Bosi, A.; Ciccola, A.; Serafini, I.; Guiso, M.; Ripanti, F.; Postorino, P.; Curini, R.; Bianco, A. Street art graffiti: Discovering their composition and alteration by FTIR and micro-Raman spectroscopy. *Spectrochim. Acta A Mol. Biomol. Spectrosc.* **2020**, *225*, 117474. [CrossRef] [PubMed]
4. Sanmartín, P.; Cappitelli, F. Evaluation of accelerated ageing tests for metallic and non-metallic graffiti paints applied to stone. *Coatings* **2017**, *7*, 180. [CrossRef]
5. Dionísio, A.; Ribeiro, T. When graffiti is not art: The damage of alkyd sprays on calcareous stones employed in Cultural Heritage. In *Cultural Heritage: Protection, Developments and International Perspectives (Focus on Civilizations and Cultures)*; Nova Science Publishers: Hauppauge, NY, USA, 2013; pp. 279–291, ISBN 978-1-62808-822-9.
6. Gomes, V.; Dionísio, A.; Pozo-Antonio, J.S. Conservation strategies against graffiti vandalism on Cultural Heritage stones: Protective coatings and cleaning methods. *Prog. Org. Coat.* **2017**, *113*, 90–109. [CrossRef]
7. Sanmartín, P.; Cappitelli, F.; Mitchell, R. Current methods of graffiti removal: A review. *Constr. Build. Mater.* **2014**, *71*, 363–374. [CrossRef]
8. Rivas, T.; Pozo, S.; Fiorucci, M.P.; López, A.J.; Ramil, A. Nd: YVO4 laser removal of graffiti from granite. Influence of paint and rock properties on cleaning efficacy. *Appl. Surf. Sci.* **2012**, *263*, 563–572. [CrossRef]
9. Samolik, S.; Walczak, M.; Plotek, M.; Sarzynski, A.; Pluska, I.; Marczak, J. Investigation into the removal of graffiti on mineral supports: Comparison of nanosecond Nd: YAG laser cleaning with traditional mechanical and chemical methods. *Stud. Conserv.* **2015**, *60*, S58–S64. [CrossRef]
10. Rossi, S.; Fedel, M.; Petrolli, S.; Deflorian, F. Behaviour of different removers on permanent anti-graffiti organic coatings. *J. Build. Eng.* **2016**, *5*, 104–113. [CrossRef]
11. Pozo-Antonio, J.S.; Rivas, T.; Fiorucci, M.P.; López, A.J.; Ramil, A. Effectiveness and harmfulness evaluation of graffiti cleaning by mechanical, chemical and laser procedures on granite. *Microchem. J.* **2016**, *125*, 1–9. [CrossRef]
12. Sanmartín, P.; Bosch-Roig, P. Biocleaning to remove graffiti: A real possibility? Advances towards a complete protocol of action. *Coatings* **2019**, *9*, 104. [CrossRef]
13. Germinario, G.; van der Werf, I.D.; Palazzo, G.; Regidor Ros, J.L.; Montes-Estelles, R.M.; Sabbatini, L. Bioremoval of marker pen inks by exploiting lipase hydrolysis. *Prog. Org. Coat.* **2017**, *110*, 162–171. [CrossRef]
14. García, O.; Malaga, K. Definition of the procedure to determine the suitability and durability of an anti-graffiti product for application on cultural heritage porous materials. *J. Cult. Herit.* **2012**, *13*, 77–82. [CrossRef]
15. Licchelli, M.; Marzolla, S.J.; Poggi, A.; Zanchi, C. Crosslinked fluorinated polyurethanes for the protection of stone surfaces from graffiti. *J. Cult. Herit.* **2011**, *12*, 34–43. [CrossRef]
16. Liu, H.; Gao, L.; Shang, Q.; Xiao, G. Preparation and characterization of polyurethane clearcoats and investigation into their antigraffiti property. *J. Coat. Technol. Res.* **2013**, *10*, 775–784. [CrossRef]
17. Rabea, A.M.; Mohseni, M.; Mirabedini, S.M.; Tabatabaei, M.H. Surface analysis and anti-graffiti behavior of a weathered polyurethane-based coating embedded with hydrophobic nano silica. *Appl. Surf. Sci.* **2012**, *258*, 4391–4396. [CrossRef]
18. Manvi, G.N.; Singh, A.R.; Jagtap, R.N.; Kothari, D.C. Isocyanurate based fluorinated polyurethane dispersion for anti-graffiti coatings. *Prog. Org. Coat.* **2012**, *75*, 139–146. [CrossRef]
19. Carmona-Quiroga, P.M.; Rubio, J.; Sánchez, M.J.; Martínez-Ramírez, S.; Blanco-Varela, M.T. Surface dispersive energy determined with IGC-ID in anti-graffiti-coated building materials. *Prog. Org. Coat.* **2011**, *71*, 207–212. [CrossRef]
20. García, O.; Rz-Maribona, I.; Gardei, A.; Riedl, M.; Vanhellemont, Y.; Santarelli, M.L.; Suput, J.S. Comparative study of the variation of the hydric properties and aspect of natural stone and brick after the application of 4 types of anti-graffiti. *Mater. Constr.* **2010**, *60*, 69–82. [CrossRef]

21. Carmona-Quiroga, P.M.; Jacobs, R.M.J.; Martínez-Ramírez, S.; Viles, H.A. Durability of anti-graffiti coatings on stone: Natural vs accelerated weathering. *PLoS ONE* **2017**, *12*, e0172347. [CrossRef]
22. Kharitonov, A.P.; Simbirtseva, G.V.; Nazarov, V.G.; Stolyarov, V.P.; Dubois, M.; Peyroux, J. Enhanced anti-graffiti or adhesion properties of polymers using versatile combination of fluorination and polymer grafting. *Prog. Org. Coat.* **2015**, *88*, 127–136. [CrossRef]
23. Esposito Corcione, C.; De Simone, N.; Santarelli, M.L.; Frigione, M. Protective properties and durability characteristics of experimental and commercial organic coatings for the preservation of porous stone. *Prog. Org. Coat.* **2017**, *103*, 193–203. [CrossRef]
24. Heinisch, M.; Miricescu, D. Innovative industrial technologies for preventive anti-graffiti coating. In *MATEC Web of Conferenes*; EDP Sciences: Les Ulis, France, 2017; Volume 121, p. 03009.
25. Licchelli, M.; Malagodi, M.; Weththimuni, M.; Zanchi, C. Anti-Graffiti nanocomposite materials for surface protection of a very porous stone. *Appl. Phys. A* **2014**, *116*, 1525–1539. [CrossRef]
26. Moura, A.; Flores-Colen, I.; De Brito, J. Study of the effect of three anti-graffiti products on the physical properties of different substrates. *Constr. Build. Mater.* **2016**, *107*, 157–164. [CrossRef]
27. Fedel, M.; Rossi, S.; Deflorian, F. Polymethyl (hydro)/polydimethylsilazane-derived coatings applied on AA1050: Effect of the dilution in butyl acetate on the structural and electrochemical properties. *J. Coat. Technol. Res.* **2019**, *16*, 1013–1019. [CrossRef]
28. Melquiades, F.L.; Appoloni, C.R.; Andrello, A.C.; Spagnuolo, E. Non-Destructive analytical techniques for the evaluation of cleaning and protection processes on white marble surfaces. *J. Cult. Herit.* **2019**, *37*, 54–62. [CrossRef]
29. Frigione, M.; Lettieri, M. Novel attribute of organic–inorganic hybrid coatings for protection and preservation of materials (stone and wood) belonging to cultural heritage. *Coatings* **2018**, *8*, 319. [CrossRef]
30. Kronlund, D.; Lindén, M.; Smått, J.-H. A sprayable protective coating for marble with water-repellent and anti-graffiti properties. *Prog. Org. Coat.* **2016**, *101*, 359–366. [CrossRef]
31. Bayer, S.I. On the durability and wear resistance of transparent superhydrophobic coatings. *Coatings* **2017**, *7*, 12. [CrossRef]
32. Bormashenko, E. Physics of solid–liquid interfaces: From the Young equation to the superhydrophobicity. *Low Temp. Phys.* **2016**, *42*, 622–635. [CrossRef]
33. Simpson, J.T.; Hunter, S.R.; Aytug, T. Superhydrophobic materials and coatings: A review. *Rep. Prog. Phys.* **2015**, *78*, 086501. [CrossRef]
34. Milionis, A.; Loth, E.; Bayer, I.S. Recent advances in the mechanical durability of superhydrophobic materials. *Adv. Colloid Interface Sci.* **2016**, *229*, 57–79. [CrossRef] [PubMed]
35. Cohen, N.; Dotan, A.; Dodiuk, H.; Kenig, S. Superhydrophobic coatings and their durability. *Mater. Manuf. Process.* **2016**, *31*, 1143–1155. [CrossRef]
36. Manoudis, P.N.; Tsakalof, A.; Karapanagiotis, I.; Zuburtikudis, I.; Panayiotou, C. Fabrication of super-hydrophobic surfaces for enhanced stone protection. *Surf. Coat. Technol.* **2009**, *203*, 1322–1328. [CrossRef]
37. Milionis, A.; Bayer, I.S.; Loth, E. Recent advances in oil-repellent surfaces. *Int. Mater. Rev.* **2016**, *61*, 101–126. [CrossRef]
38. Hosseini, M.; Karapanagiotis, I. (Eds.) *Advanced Materials for the Conservation of Stone*; Springer: Cham, Switzerland, 2018; ISBN 978-3-319-72260-3.
39. Godeau, G.; Guittard, F.; Darmanin, T. Surfaces bearing fluorinated nucleoperfluorolipids for potential anti-graffiti surface properties. *Coatings* **2017**, *7*, 220. [CrossRef]
40. Haas, K.-H.; Amberg-Schwab, S.; Rose, K. Functionalized coating materials based on inorganic-organic polymers. *Thin Solid Film.* **1999**, *351*, 198–203. [CrossRef]
41. Badila, M.; Kohlmayr, M.; Zikulnig-Rusch, E.M.; Dolezel-Horwath, E.; Kandelbauer, A. Improving the cleanability of melamine-formaldehyde-based decorative laminates. *J. Appl. Polym. Sci.* **2014**, *131*, 40964. [CrossRef]
42. Xu, F.; Li, X.; Li, Y.; Sun, J. Oil-repellent antifogging films with water-enabled functional and structural healing ability. *ACS Appl. Mater. Interfaces* **2017**, *9*, 27955–27963. [CrossRef]
43. Carmona-Quiroga, P.M.; Martínez-Ramírez, S.; Sánchez-Cortés, S.; Oujja, M.; Castillejo, M.; Blanco-Varela, M.T. Effectiveness of antigraffiti treatments in connection with penetration depth determined by different techniques. *J. Cult. Herit.* **2010**, *11*, 297–303. [CrossRef]

44. Khan, F.; Khan, A.; Tuhin, M.O.; Rabnawaz, M.; Li, Z.; Naveed, M. A novel dual-layer approach towards omniphobic polyurethane coatings. *RSC Adv.* **2019**, *9*, 26703–26711. [CrossRef]
45. Malaga, K.; Mueller, U. Relevance of hydrophobic and oleophobic properties of antigraffiti systems on their cleaning efficiency on concrete and stone surfaces. *J. Mater. Civ. Eng.* **2013**, *25*, 755–762. [CrossRef]
46. Scheerder, J.; Visscher, N.; Nabuurs, T.; Overbeek, A. Novel, water-based fluorinated polymers with excellent antigraffiti properties. *J. Coat. Technol. Res.* **2005**, *2*, 617–625. [CrossRef]
47. Lettieri, M.; Masieri, M.; Morelli, A.; Pipoli, M.; Frigione, M. Oleo/hydrophobic coatings containing nano-particles for the protection of stone materials having different porosity. *Coatings* **2018**, *8*, 429. [CrossRef]
48. Bugani, S.; Camaiti, M.; Morselli, L.; Van de Casteele, E.; Janssens, K. Investigation on porosity changes of Lecce stone due to conservation treatments by means of X-ray nano- and improved micro-computed tomography: Preliminary results. *X-Ray Spectrom.* **2007**, *36*, 316–320. [CrossRef]
49. Föllmi, K.B.; Hofmann, H.; Chiaradia, M.; de Kaenel, E.; Frijia, G.; Parente, M. Miocene phosphate-rich sediments in Salento (southern Italy). *Sediment. Geol.* **2015**, *327*, 55–71. [CrossRef]
50. Tiano, P.; Accolla, P.; Tomaselli, L. Phototrophic biodeteriogens on lithoid surfaces: An ecological study. *Microb. Ecol.* **1995**, *29*, 299–309. [CrossRef] [PubMed]
51. Dunham, R.J. Classification of carbonate rocks according to depositional textures. *AAPG Mem.* **1962**, *1*, 108–121.
52. NORMAL Rec. 4/80 *Distribuzione del Volume dei Pori in Funzione del Loro Diametro*; CNR/ICR: Rome, Italy, 1980.
53. UNI 10921. *Beni Culturali Materiali Lapidei Naturali ed Artificiali—Prodotti Idrorepellenti—Applicazione su Provini e Determinazione in Laboratorio Delle Loro Caratteristiche*; Ente Italiano di normazione: Milan, Italy, 2001.
54. *ASTM D 6578. Standard Practice for Determination of Graffiti Resistance*; ASTM International: West Conshohocken, PA, USA, 2000.
55. Baglioni, M.; Poggi, G.; Jaidar Benavides, Y.; Martínez Camacho, F.; Giorgi, R.; Baglioni, P. Nanostructured fluids for the removal of graffiti—A survey on 17 commercial spray-can paints. *J. Cult. Herit.* **2018**, *34*, 218–226. [CrossRef]
56. EN 15802. *Conservation of Cultural Property—Test Methods—Determination of Static Contact Angle*; CEN (European Committee for Standardization): Brussels, Belgium, 2010.
57. Aslanidou, D.; Karapanagiotis, I.; Panayiotou, C. Tuning the wetting properties of siloxane-nanoparticle coatings to induce superhydrophobicity and superoleophobicity for stone protection. *Mater. Des.* **2016**, *108*, 736–744. [CrossRef]
58. Facio, D.S.; Carrascosa, L.A.M.; Mosquera, M.J. Producing lasting amphiphobic building surfaces with self-cleaning properties. *Nanotechnology* **2017**, *28*, 265601. [CrossRef]
59. Aslanidou, D.; Karapanagiotis, I.; Lampakis, D. Waterborne superhydrophobic and superoleophobic coatings for the protection of marble and sandstone. *Materials* **2018**, *11*, 585. [CrossRef] [PubMed]
60. EN 15886. *Conservation of Cultural Property–Test Methods—Colour Measurement of Surfaces*; CEN (European Committee for Standardization): Brussels, Belgium, 2010.
61. Sahasrabudhe, S.N.; Rodriguez-Martinez, V.; O'Meara, M.; Farkas, B.E. Density, viscosity, and surface tension of five vegetable oils at elevated temperatures: Measurement and modeling. *Int. J. Food Prop.* **2017**, *20*, 1965–1981. [CrossRef]
62. Rabea, A.M.; Mirabedini, S.M.; Mohseni, M. Investigating the surface properties of polyurethane based anti-graffiti coatings against UV exposure. *J. Appl. Polym. Sci.* **2012**, *124*, 3082–3091. [CrossRef]
63. Shang, B.; Chen, M.; Wu, L. One-step synthesis of statically amphiphilic/dynamically amphiphobic fluoride-free transparent coatings. *ACS Appl. Mater. Interfaces* **2018**, *10*, 41824–41830. [CrossRef] [PubMed]
64. Zhong, X.; Hu, H.; Yang, L.; Sheng, J.; Fu, H. Robust hyperbranched polyester-based anti-smudge coatings for self-cleaning, anti-graffiti, and chemical shielding. *ACS Appl. Mater. Interfaces* **2019**, *11*, 14305–14312. [CrossRef] [PubMed]
65. Carvalhão, M.; Dionísio, A. Evaluation of mechanical soft-abrasive blasting and chemical cleaning methods on alkyd-paint graffiti made on calcareous stones. *J. Cult. Herit.* **2015**, *16*, 579–590. [CrossRef]
66. Gherardi, F.; Colombo, A.; D'Arienzo, M.; Di Credico, B.; Goidanich, S.; Morazzoni, F.; Simonutti, R.; Toniolo, L. Efficient self-cleaning treatments for built heritage based on highly photo-active and well-dispersible TiO_2 nanocrystals. *Microchem. J.* **2016**, *126*, 54–62. [CrossRef]
67. Gomes, V.; Dionísio, A.; Pozo-Antonio, J.S. The influence of the SO_2 ageing on the graffiti cleaning effectiveness with chemical procedures on a granite substrate. *Sci. Total Environ.* **2018**, *625*, 233–245. [CrossRef]

68. Masieri, M.; Lettieri, M. Influence of the distribution of a spray paint on the efficacy of anti-graffiti coatings on a highly porous natural stone material. *Coatings* **2017**, *7*, 18. [CrossRef]
69. Tarnowski, A.; Zhang, X.; Mcnamara, C.; Martin, S.; Mitchell, R. Biodeterioration and performance of anti-graffiti coatings on sandstone and marble. *J. Can. Assoc. Conserv. J CAC* **2007**, *32*, 3–16.
70. Pozo-Antonio, J.S.; Rivas, T.; Jacobs, R.M.J.; Viles, H.A.; Carmona-Quiroga, P.M. Effectiveness of commercial anti-graffiti treatments in two granites of different texture and mineralogy. *Prog. Org. Coat.* **2018**, *116*, 70–82. [CrossRef]
71. Poli, T.; Toniolo, L. The challenge of protecting monuments from atmospheric attack. In *Fracture and Failure of Natural Building Stones: Applications in the Restoration of Ancient Monuments*; Kourkoulis, S.K., Ed.; Springer Netherlands: Dordrecht, The Netherlands, 2006; pp. 553–563, ISBN 978-1-4020-5077-0.
72. Selwitz, C. *Epoxy Resins in Stone Conservation*; Getty Publications: Marina del Rey, CA, USA, 1992; Volume 7, ISBN 0-89236-238-3.
73. Sena da Fonseca, B.; Piçarra, S.; Ferreira Pinto, A.P.; Ferreira, M.J.; Montemor, M.F. TEOS-based consolidants for carbonate stones: The role of N1-(3-trimethoxysilylpropyl)diethylenetriamine. *New J. Chem.* **2017**, *41*, 2458–2467. [CrossRef]
74. Borsoi, G.; Lubelli, B.; van Hees, R.; Veiga, R.; Silva, A.S.; Colla, L.; Fedele, L.; Tomasin, P. Effect of solvent on nanolime transport within limestone: How to improve in-depth deposition. *Colloids Surf. Physicochem. Eng. Asp.* **2016**, *497*, 171–181. [CrossRef]
75. Domaslowski, W. The mechanism of polymer migration in porous stones. *Wien. Ber. Über Nat. Kunst* **1988**, *4*, 402–425.
76. Hansen, E.F.; Lowinger, R.; Sadoff, E. Consolidation of porous paint in a vapor-saturated atmosphere a technique for minimizing changes in the appearance of powdering, matte paint. *J. Am. Inst. Conserv.* **1993**, *32*, 1–14.
77. Sena da Fonseca, B.; Piçarra, S.; Pinto, A.P.; Montemor, M.d.F. Polyethylene glycol oligomers as siloxane modificators in consolidation of carbonate stones. *Pure Appl. Chem.* **2016**, *88*, 1117. [CrossRef]
78. Giorgiutti-Dauphiné, F.; Pauchard, L. Painting cracks: A way to investigate the pictorial matter. *J. Appl. Phys.* **2016**, *120*, 065107. [CrossRef]
79. Krzemień, L.; Łukomski, M.; Bratasz, Ł.; Kozłowski, R.; Mecklenburg, M.F. Mechanism of craquelure pattern formation on panel paintings. *Stud. Conserv.* **2016**, *61*, 324–330. [CrossRef]
80. Atkinson, A.; Guppy, R.M. Mechanical stability of sol-gel films. *J. Mater. Sci.* **1991**, *26*, 3869–3873. [CrossRef]
81. Moura, A.R.; Flores-Colen, I.; de Brito, J. Anti-Graffiti products for porous surfaces. An overview. In Proceedings of the Hydrophobe VII 7th International Conference on Water Repellent Treatment and Protective Surface Technology for Building Materials, LNEC (Laboratório Nacional de Engenharia Civil), Lisbon, Portugal, 11–12 September 2014; pp. 225–233.
82. Moretti, P.; Germinario, G.; Doherty, B.; van der Werf, I.D.; Sabbatini, L.; Mirabile, A.; Sgamellotti, A.; Miliani, C. Disclosing the composition of historical commercial felt-tip pens used in art by integrated vibrational spectroscopy and pyrolysis-gas chromatography/mass spectrometry. *J. Cult. Herit.* **2019**, *35*, 242–253. [CrossRef]
83. Lettieri, M.; Masieri, M. Surface characterization and effectiveness evaluation of anti-graffiti coatings on highly porous stone materials. *Appl. Surf. Sci.* **2014**, *288*, 466–477. [CrossRef]

© 2019 by the authors. Licensee MDPI, Basel, Switzerland. This article is an open access article distributed under the terms and conditions of the Creative Commons Attribution (CC BY) license (http://creativecommons.org/licenses/by/4.0/).

Article

A Facile Route to Fabricate Superhydrophobic Cu₂O Surface for Efficient Oil–Water Separation

Sheng Lei [1], Xinzuo Fang [1], Fajun Wang [1,2,*], Mingshan Xue [1,2,*], Junfei Ou [1,2], Changquan Li [1] and Wen Li [1]

- [1] School of Materials Engineering, Jiangsu University of Technology, Changzhou 213001, China; shenglei@jsut.edu.cn (S.L.); fangxz@mail.ustc.edu.cn (X.F.); oujunfei_1982@163.com (J.O.); 70225@nchu.edu.cn (C.L.); 2018500123@jsut.edu.cn (W.L.)
- [2] School of Materials Science and Engineering, Nanchang Hangkong University, Nanchang 330063, China
- * Correspondence: jjbxsjz@foxmails.com (F.W.); xuems04@mails.ucas.ac.cn (M.X.)

Received: 18 September 2019; Accepted: 11 October 2019; Published: 12 October 2019

Abstract: The mixture of insoluble organics and water seriously affects human health and environmental safety. Therefore, it is important to develop an efficient material to remove oil from water. In this work, we report a superhydrophobic Cu_2O mesh that can effectively separate oil and water. The superhydrophobic Cu_2O surface was fabricated by a facile chemical reaction between copper mesh and hydrogen peroxide solution without any low surface reagents treatment. With the advantages of simple operation, short reaction time, and low cost, the as-synthesized superhydrophobic Cu_2O mesh has excellent oil–water selectivity for many insoluble organic solvents. In addition, it could be reused for oil–water separation with a high separation ability of above 95%, which demonstrated excellent durability and reusability. We expect that this fabrication technique will have great application prospects in the application of oil–water separation.

Keywords: superhydrophobic; Cu_2O; oil–water separation

1. Introduction

Oil in water can reduce the purity of water, and the oily wastewater generated in daily life and industry can also cause serious pollution to the environment [1,2]. Moreover, the presence of water in oil can seriously affect the quality and efficiency of oil such as reducing the service efficiency and life of an engine. Therefore, the research of the separation of the oil–water mixture is of great significance and has broad application prospects. Meanwhile, efficient oil–water separation technology has also attracted great attention [3–7]. In 2004, Jiang et al. prepared a superhydrophobic and superoleophilic coating mesh by spraying polytetrafluoroethylene onto the stainless steel mesh. The contact angles of water and diesel oil on this mesh were 156.2° and 0°, respectively, which realized the effective separation of diesel and water [8]. Inspired by this, many scientists have developed great interest in the application of special wettability materials in oil–water separation.

Over the past decades, researchers have produced a wide variety of materials with superhydrophobic and superoleophilic properties by manipulating and modifying the surface chemical composition and surface roughness. Examples include superhydrophobic metal mesh [9–12], polyurethane sponge [13–15], fiber textile [16–19], metal foam [20,21], polymer membranes [22–24], and so on [25–27], which can selectively repel water from the mixtures of oil and water while allowing oil to penetrate through the materials or to be absorbed, and exhibit high efficiency of oil–water separation performance. However, the preparation process of most of the superhydrophobic materials above-mentioned is complicated and time-consuming, requires special equipment and costly reagents, which severely limits their large-scale production and practical application. Moreover, most superhydrophobic materials may contaminate oil during oil–water separation due to the use of low surface energy reagents that are not conducive to human health. Therefore, it is reasonably important to develop superhydrophobic oil–water separation materials that are environmentally friendly and without low surface energy reagents.

In this paper, the preparation of a Cu_2O film on copper mesh substrates by a simple one-step chemical reaction method is reported. The as-prepared flower-tufted Cu_2O nanosheet film exhibited superhydrophobicity without modification with low surface energy reagents, as Cu_2O is one of the rare materials that possess a low surface energy and rough structure [28–31]. Water has weak interaction with Cu_2O and does not form bonds on its surface, resulting in the hydrophobicity of Cu_2O. Moreover, the relationship between the reaction time and growth process, morphology, and hydrophobicity of Cu_2O film were investigated. The superhydrophobic Cu_2O mesh also exhibited a superoleophilic property and successfully achieved the separation of various insoluble oil–water mixtures, demonstrating high separation efficiency and reusability. Although Cu_2O formation has been reported by previous papers, a simpler method for preparing superhydrophobic Cu_2O still needs to be developed. Therefore, a method for preparing Cu_2O by reacting copper mesh with hydrogen peroxide is more industrially practical. In addition, the superhydrophobic Cu_2O mesh without similar fluorine-containing reagents is more beneficial to human health, especially for oil–water separation in the production process of edible oils.

2. Materials and Methods

The red copper mesh (purity ≥99.97%, 200) was purchased from Anping Tairun Wire Mesh Co. Ltd., Hengshui, China. Hydrogen peroxide solution (H_2O_2, 30 wt % in H_2O), hydrochloric acid, absolute ethyl alcohol, and acetone were bought from Shanghai Chemical Reagent Co. Ltd., Shanghai, China. All chemical reagents used were analytical grade and did not require further processing.

The copper mesh of 30 mm × 30 mm was ultrasonically cleaned with 0.1 M HCl solution, acetone, and deionized water for 2 min before use, respectively. Then, it was immersed in 100 mL of hydrogen peroxide solution for reaction. After a period of reaction, the copper mesh was taken out and cleaned successively with deionized water and ethanol. Finally, the copper mesh was dried at 120 °C for 5 h under an air atmosphere.

The surface morphology of the copper mesh was determined by a field emission scanning electron microscope (SEM, Sigma 500, Zeiss, Oberkochen, Germany) instrument at 5–15 kV under a vacuum environment. The surface chemical compositions were untreated and the fabricated meshes were measured by a PHI-5702 X-ray photoelectron spectroscopy (XPS, Kratos Analytical Ltd., Manchester, UK). Water contact angle (WCA), oil contact angle (OCA), and water sliding angle (WSA) were measured using an optical contact angle meter (DSA 30, Krüss, Hamburg, Germany) with 5 µL droplets at ambient temperature. The average WCA, OCA, and WSA values were obtained by measuring the same sample in at least five different positions.

3. Results and Discussion

The surface chemical composition of the bare copper mesh and the as-prepared superhydrophobic copper mesh were analyzed in detail using XPS, as depicted in Figure 1. There were three main elements of Cu, O, and C in the XPS survey spectrum of the bare and the superhydrophobic copper mesh surface in Figure 1a. In the high-resolution XPS spectra of the bare copper mesh (Figure 1b), it exhibited two strong peaks centered at 952.4 and 932.6 eV, corresponding to the Cu double peaks of Cu $2p_{1/2}$ and Cu $2p_{3/2}$, respectively [32–35]. Figure 1c shows the photoelectron spectrum of the Cu $2p$ core level for the superhydrophobic copper mesh. Two peaks located at the binding energies of 952.5 and 932.5 eV can be attributed to Cu $2p_{1/2}$ and Cu $2p_{3/2}$, respectively, which agreed well with Cu_2O [36–40]. The O $1s$ spectrum of the superhydrophobic copper mesh in Figure 1d showed only one peak at 531.1 eV, which can be attributed to Cu_2O [36,39,40]. The above results confirm that the surface of the superhydrophobic copper mesh is mainly Cu_2O. Therefore, it is presumed that hydrogen peroxide reacts with the copper mesh, leading to the conversion of the upmost layer of Cu to Cu_2O, according to the following reactions:

$$2Cu + H_2O_2 \rightarrow Cu_2O + H_2O \tag{1}$$

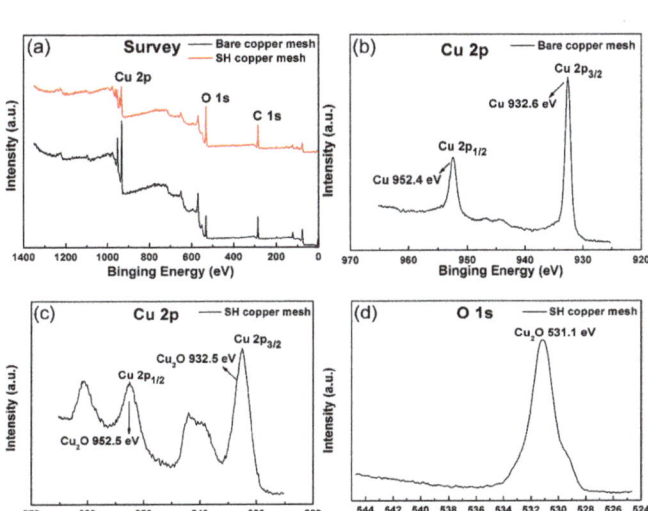

Figure 1. The XPS spectra of the bare and superhydrophobic copper mesh: survey (**a**), Cu 2*p* of the bare copper mesh (**b**), Cu 2*p* (**c**) and O 1*s* (**d**) of the superhydrophobic copper mesh.

The surface microstructure was considered to be a key factor affecting the surface superhydrophobic properties of the material surface. In the experiment, keeping the concentration and volume of the hydrogen peroxide solution constant, the reaction was carried out by changing the time. The growth process and morphology of the Cu_2O film that changed with reaction time are presented in Figure 2. When the reaction time was 10 min, a large amount of worm-like Cu_2O nanosheets were generated on the surface of the copper mesh (Figure 2a). However, careful observation showed that there were also many Cu_2O nanoparticles presented on the copper mesh surface. After 20 min of reaction, the worm-like nanosheets began to form numerous honeycomb nanosheets. The surface of the copper mesh was mainly covered by honeycomb Cu_2O nanosheet structures and grooves (Figure 2b). After reacting for 30 min, the trend of the growth of the Cu_2O nanosheets became very obvious, as shown in Figure 2c. The surface of the copper mesh was completely covered by a large number of flower-tufted nanosheet structures, which were dense, thin, and relatively uniform in size. Prolonging the reaction time to 40 min, the size of the Cu_2O nanosheets was enlarged and appeared sparse, but the lamellar structure was relatively regular and orderly, as shown in Figure 2d. Increasing the reaction time to 60 min, the Cu_2O nanosheet structure on the surface of the copper mesh showed disorder (Figure 2e). The reason may be that the heat released in the reaction process causes the solution temperature to rise and the reaction to intensify, resulting in corrosion of the partially Cu_2O nanosheet structure. With a further increase of the reaction time to 120 min, the surface of the copper mesh was covered with leaf-like Cu_2O nanosheets, which were thick and uneven in size, as shown in Figure 2f. It is possible that Cu_2O restored the growth of the nanosheet structure due to the decrease in the hydrogen peroxide concentration after a long period of reaction.

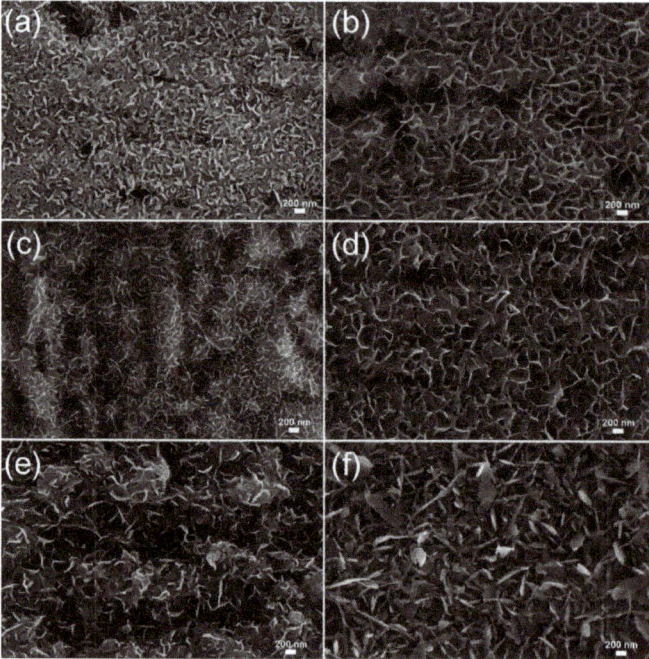

Figure 2. SEM images of the as-prepared Cu$_2$O surface on copper mesh with different reaction times: (**a**) 10 min, (**b**) 20 min, (**c**) 30 min, (**d**) 40 min, (**e**) 60 min, and (**f**) 120 min.

In order to investigate the effect of the reaction time on wettability, the water contact angles and sliding angles of the Cu$_2$O surface prepared at different times were detected and are shown in Figure 3. When the reaction time was 10 min, it can be clearly seen that the water contact angle was 148.7° and the sliding angle was 69.8°. When reacted for 20 min, the contact angle reached 157.2°, while the sliding angle was quickly reduced to 17.5°. Prolonging the reaction time to 30 min, the water contact angle and sliding angle were 165.4° and 5.6°, respectively. At this time, the prepared Cu$_2$O surface exhibited superhydrophobicity. Moreover, when the reaction time increased from 40 to 120 min, the water contact angle and sliding angle of the Cu$_2$O surface slightly increased and decreased, respectively. This demonstrated that all of the Cu$_2$O surfaces prepared after 30 min of reaction had good superhydrophobicity and low adhesive hydrophobicity. Therefore, the superhydrophobic Cu$_2$O surface obtained in 30 min was the best in terms of cost and time. Test results shown in Figure 3 also indicated that the hydrophobic property of the prepared Cu$_2$O surface increased with the extension of reaction time. On the one hand, the amount of Cu$_2$O increased with the reaction time, which gradually covered the surface of the copper mesh and reduced its surface energy. On the other hand, the Cu$_2$O nanosheet structures of the copper mesh surface changed with the reaction time, and the corresponding morphology is shown in Figure 2. These micro/nano-rough structures contribute to reduce the surface contact area with water droplets.

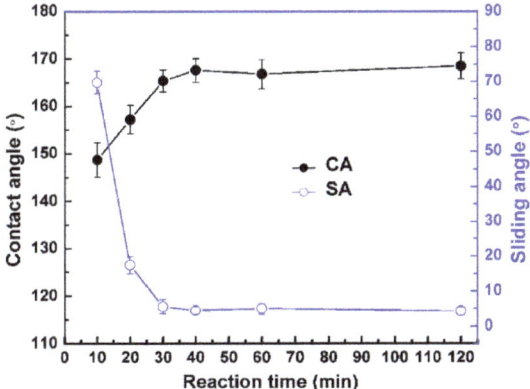

Figure 3. Water contact angles and sliding angles of the as-prepared Cu$_2$O surface with different reaction times.

Figure 4 displays the surface morphology and water wetting properties of the bare copper mesh and superhydrophobic Cu$_2$O mesh. As illustrated in Figure 4a, the surface of the bare copper mesh was smooth, which could be clearly observed even in the SEM image at higher magnification (Figure 4b). The contact angle of the water droplets on the surface of the copper mesh was 119.8°, but the water droplets did not drop when rotated 90°. In addition, the contact angle of organic solvents such as toluene, trichloromethane, gasoline, and kerosene on the surface of the bare copper mesh was close to 0°. As evidenced from Figure 4c, the superhydrophobic Cu$_2$O mesh surface was very rough, which was prepared by reacting for 30 min. In the high-magnification SEM image, many flower-tufted Cu$_2$O nanosheets were observed on the surface of the copper mesh (Figure 4d). The contact angle of the water droplets on the surface of the superhydrophobic Cu$_2$O mesh reached 165.4°, and the sliding angle was as low as 5.6°. However, the contact angle of toluene, trichloromethane, and other oils in the superhydrophobic Cu$_2$O mesh was approximately 0°, which confirmed that the prepared superhydrophobic Cu$_2$O mesh also had excellent superoleophilicity. The pore size of the copper mesh was about 75 μm, and the hierarchical micro/nanostructure was constructed by combining the flower-tufted Cu$_2$O nanosheets on the surface. The grooves and gaps created by these rough structures can capture a large volume of air, resulting in large water contact angles and small sliding angles on the surface, which can be explained by the Cassie–Baxter equation [41]:

$$\cos \theta r = f_1 \cos \theta - f_2 \quad (2)$$

where θr is the contact angle of liquid on the rough surface; θ is the contact angle of liquid on the corresponding smooth surface; f_1 is the proportion of the solid surface actually in contact with liquid; f_2 is the proportion of air trapped in the hole in contact with liquid; and $f_1 + f_2 = 1$. The water contact angles on the rough superhydrophobic Cu$_2$O mesh and the smooth copper mesh were 165.4° and 119.8°, respectively. Therefore, the f_2 of the superhydrophobic Cu$_2$O mesh calculated by Equation (2) was 0.936, indicating that the contact area between water and air accounted for up to 93.6%. Correspondingly, the contact between water and solid surface only accounts for 6.4%, which is similar to the results reported by Karapanagiotis et al. [42]. The small water-solid contact area demonstrates the good superhydrophobic property on the surface.

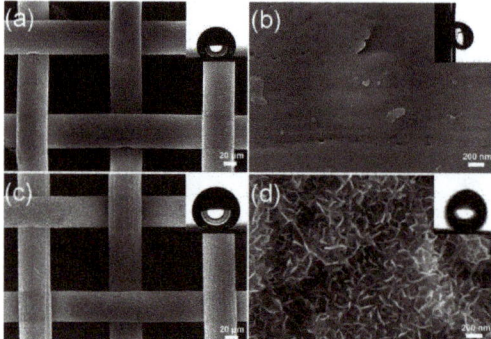

Figure 4. SEM images of the surfaces of bare (**a,b**) and superhydrophobic Cu_2O mesh (**c,d**). The insets show the optical images of the water contact angle and sliding angle.

The oil–water separation property of the as-prepared superhydrophobic Cu_2O mesh was investigated with a self-made device, as shown in Figure 5. The filter was made of superhydrophobic Cu_2O mesh sandwiched between the upper glass cylinder and the lower funnel. The superhydrophobic mesh had a thickness of 0.1 mm and was firmly fixed, so that water did not leak from the contact surface. The effective filter diameter of the filter device was 20 mm. When a drop of toluene dyed with oil red O contacted the surface of the superhydrophobic Cu_2O mesh, it spread rapidly and penetrated the mesh, leaving only red marks on the underlying filter paper. However, the water droplet remained spherical on the surface of the superhydrophobic Cu_2O mesh (Figure 5a). A total of 10 mL toluene and 10 mL water dyed with methylene blue were mixed into a 20 mL insoluble oil–water mixture (Figure 5b). As the oil–water mixture was slowly injected into the separation device, the oil easily passed through the superhydrophobic Cu_2O mesh into the beaker under the action of gravity (Figure 5c). After the oil–water separation was completed, the oil was collected in the beaker while the water was kept in the container on the superhydrophobic Cu_2O mesh (Figure 5d). The same separation process was also applicable to the mixture of water and other oils such as gasoline, kerosene, trichloromethane, hexane, edible oil, and so on. In addition, the oil flux of the superhydrophobic Cu_2O mesh was measured and the values were calculated by the following equation:

$$J = V/St \qquad (3)$$

where V is the volume of oil is 0.01 L; S is the effective oil-passing area of the superhydrophobic Cu_2O mesh; and t is the time for the permeation of 0.01 L of oil. As a result, the flux values of toluene, gasoline, kerosene, trichloromethane, hexane, and edible oil of the superhydrophobic Cu_2O mesh were 8.18, 7.68, 7.72, 8.36, 8.12, and 0.53 $L \cdot m^{-2} \cdot s^{-1}$, respectively. The viscosity of edible oil was larger than that of other oils, which led to a long time required to penetrate the copper mesh, and the flux value was small.

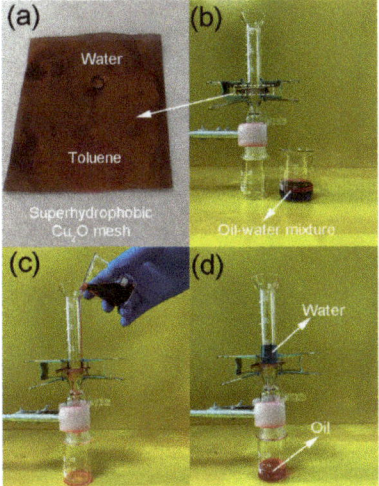

Figure 5. (**a**) Water and oil on the surface of the superhydrophobic Cu$_2$O mesh, (**b**) separation device and oil–water mixture (toluene was dyed with oil red O and water was dyed with methylene blue), (**c**,**d**) oil–water separation process.

The oil–water separation performance was investigated using the as-prepared superhydrophobic Cu$_2$O mesh. The oil–water separation efficiency of the superhydrophobic Cu$_2$O mesh was evaluated and analyzed by the following formula:

$$\eta = V/V_0 \times 100\% \tag{4}$$

where η represents the separation efficiency and V and V_0 are the oil volume before and after the separation experiment, respectively [43,44]. Figure 6a shows the separation efficiency of several different types of oil and water mixtures. It can be clearly seen that the initial separation efficiency of trichloromethane, toluene, gasoline, kerosene, hexane, and edible oil all exceeded 95%. The relatively low separation efficiency of edible oil was mainly due to its high viscosity, which was slightly stuck to the container wall and mesh surface during the oil–water separation process. More importantly, the superhydrophobic Cu$_2$O mesh exhibited excellent reusability. As illustrated in Figure 6b, the oil–water separation efficiency was still as high as 96.9% after 50 cycles of repeated separation test of the toluene and water mixture. The slight decrease in the oil–water separation efficiency could be attributed to the volatilization and loss of toluene in the test process. In addition, the superhydrophobic Cu$_2$O mesh had the same superhydrophobicity and superoleophilicity after washing and drying, indicating its good potential application prospects in oil–water separation.

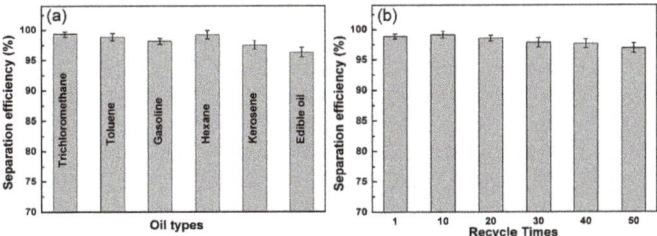

Figure 6. Oil separation efficiency of the superhydrophobic Cu$_2$O mesh: (**a**) for various oil–water mixtures in the first cycle, (**b**) for toluene after different separation cycles.

4. Conclusions

In summary, the superhydrophobic Cu_2O mesh was prepared by a one-step reaction of red copper mesh and hydrogen peroxide without modification of any low surface energy reagent. The growth process, morphology, and hydrophobicity of the Cu_2O film changed with the increase in reaction time. The superhydrophobic and superoleophilic Cu_2O mesh with the flower-tufted nanosheet structures was obtained at an optimum reaction time of 30 min, and the water contact angle, sliding angle, and oil contact angle were 165.4°, 5.6°, and 0°, respectively. Moreover, the as-prepared superhydrophobic Cu_2O mesh could separate various insoluble oil–water mixtures, and the oil–water separation efficiency was more than 95%. In addition, the superhydrophobic Cu_2O mesh displayed good reusability and stability after 50 oil–water separation cycles. This simple, low-cost, large-area preparation technique for superhydrophobic Cu_2O mesh will have a wide application prospect in the field of oil–water separation.

Author Contributions: Conceptualization, S.L.; Methodology, S.L.; Resources, S.L., F.W., and M.X.; Validation, X.F., M.X., and J.O.; Investigation, S.L., X.F., and J.O.; Data curation, S.L.; Writing-Original Draft Preparation, S.L.; Writing-Review and Editing, S.L., F.W., and M.X.; Supervision, C.L.; Project Administration, S.L.; Funding Acquisition, M.X. and W.L.

Funding: The work was supported by the National Natural Science Foundation of China (Nos. 51662032 and 11864024), the Natural Science Foundation of Department of Science and Technology of Jiangsu Province (No. BK20191478), the Innovative and Entrepreneurial Talent Plan of Jiangsu Province, and the Qing Lan Project of Jiangsu Province.

Conflicts of Interest: The authors declare no conflicts of interest.

References

1. Ahmed, A.F.; Ahmad, J.; Basma, Y.; Ramzi, T. Assessment of alternative management techniques of tank bottom petroleum sludge in Oman. *J. Hazard. Mater.* **2007**, *141*, 557–564.
2. Guterman, L. Exxon valdez turns 20. *Science* **2009**, *323*, 1558–1559. [CrossRef] [PubMed]
3. Chen, G.H.; He, G.H. Separation of water and oil from water-in-oil emulsion by freeze/thaw method. *Sep. Purif. Technol.* **2003**, *31*, 83–89. [CrossRef]
4. Mullin, J.V.; Champ, M.A. Introduction/overview to in situ burning of oil spills. *Spill Sci. Technol. Bull.* **2003**, *8*, 323–330. [CrossRef]
5. Zouboulis, A.I.; Avranas, A. Treatment of oil-in-water emulsions by coagulation and dissolved-air flotation. *Colloid. Surface A* **2000**, *172*, 153–161. [CrossRef]
6. Adebajo, M.O.; Frost, R.L.; Kloprogge, J.T.; Carmody, O.; Kokot, S. Porous materials for oil spill cleanup: a review of synthesis and absorbing properties. *J. Porous Mater.* **2003**, *10*, 159–170. [CrossRef]
7. Nyankson, E.; Olasehinde, O.; John, V.T.; Gupta, R.B. Surfactant-loaded halloysite clay nanotube dispersants for crude oil spill remediation. *Ind. Eng. Chem. Res.* **2015**, *54*, 9328–9341. [CrossRef]
8. Feng, L.; Zhang, Z.Y.; Mai, Z.H.; Ma, Y.M.; Liu, B.Q.; Jiang, L.; Zhu, D.B. A super-hydrophobic and super-oleophilic coating mesh film for the separation of oil and water. *Angew. Chem. Int. Ed.* **2004**, *43*, 2012–2014. [CrossRef]
9. Gao, Y.F.; Cheng, M.J.; Wang, B.L.; Feng, Z.G.; Shi, F. Diving-surfacing cycle within a stimulus-responsive smart device towards developing functionally cooperating systems. *Adv. Mater.* **2010**, *22*, 5125–5128. [CrossRef]
10. Pan, Q.M.; Wang, M. Miniature boats with striking loading capacity fabricated from superhydrophobic copper meshes. *ACS Appl. Mater. Inter.* **2009**, *1*, 420–423. [CrossRef]
11. Jiang, Z.X.; Geng, L.; Huan, Y.D. Design and fabrication of hydrophobic copper mesh with striking loading capacity and pressure resistance. *J. Phys. Chem. C* **2010**, *114*, 9370–9378. [CrossRef]
12. Tian, D.L.; Zhang, X.F.; Zhai, J.; Jiang, L. Photocontrollable water permeation on the micro/nanoscale hierarchical structured ZnO mesh films. *Langmuir* **2011**, *27*, 4265–4270. [CrossRef] [PubMed]
13. Zhu, Q.; Pan, Q.M.; Liu, F.T. Facile removal and collection of oils from water surfaces through superhydrophobic and superoleophilic sponges. *J. Phys. Chem. C* **2011**, *115*, 17464–17470. [CrossRef]
14. Zhou, X.Y.; Zhang, Z.Z.; Xu, X.H.; Men, X.H.; Zhu, X.T. Facile fabrication of superhydrophobic sponge with selective sbsorption and collection of oil from water. *Ind. Eng. Chem. Res.* **2013**, *52*, 9411–9416. [CrossRef]

15. Wang, C.F.; Lin, S.J. Robust superhydrophobic/superoleophilic sponge for effective continuous absorption and expulsion of oil pollutants from water. *ACS Appl. Mater. Inter.* **2013**, *5*, 8861–8864. [CrossRef]
16. Li, S.H.; Huang, J.Y.; Chen, Z.; Chen, G.Q.; Lai, Y.K. A review on special wettability textiles: theoretical models, fabrication technologies and multifunctional applications. *J. Mater. Chem. A* **2017**, *5*, 31–55. [CrossRef]
17. Xu, Z.G.; Zhao, Y.; Wang, H.X.; Zhou, H.; Qin, C.X.; Wang, X.G.; Lin, T. Fluorine-free superhydrophobic coatings with pH-induced wettability transition for controllable oil-water separation. *ACS Appl. Mater. Inter.* **2016**, *8*, 5661–5667. [CrossRef]
18. Zhang, M.; Li, J.; Zang, D.L.; Lu, Y.; Gao, Z.X.; Shi, J.Y.; Wang, C.Y. Preparation and characterization of cotton fabric with potential use in UV resistance and oil reclaim. *Carbohyd. Polym.* **2016**, *137*, 264–270. [CrossRef]
19. Ou, J.F.; Wan, B.B.; Wang, F.J.; Xue, M.S.; Wu, H.M.; Li, W. Superhydrophobic fibers from cigarette filters for oil spill cleanup. *RSC Adv.* **2016**, *6*, 44469–44474. [CrossRef]
20. Cheng, M.J.; Gao, Y.F.; Guo, X.P.; Shi, Z.Y.; Chen, J.F.; Shi, F. A functionally integrated device for effective and facile oil spill cleanup. *Langmuir* **2011**, *27*, 7371–7375. [CrossRef]
21. Jin, X.; Shi, B.R.; Zheng, L.C.; Pei, X.H.; Zhang, X.Y.; Sun, Z.Q.; Du, Y.; Kim, J.H.; Wang, X.L.; Dou, S.X.; et al. Bio-inspired multifunctional metallic foams through the fusion of different biological solutions. *Adv. Funct. Mater.* **2014**, *24*, 2721–2726. [CrossRef]
22. Zhang, W.; Shi, Z.; Zhang, F.; Liu, X.; Jin, J.; Jiang, L. Superhydrophobic and superoleophilic PVDF membranes for effective separation of water-in-oil emulsions with high flux. *Adv. Mater.* **2013**, *25*, 2071–2076. [CrossRef] [PubMed]
23. Huang, M.L.; Si, Y.; Tang, X.M.; Zhu, Z.G.; Ding, B.; Liu, L.F.; Zheng, G.; Luo, W.J.; Yu, J.Y. Gravity driven separation of emulsified oil-water mixtures utilizing in situ polymerized superhydrophobic and superoleophilic nanofibrous membranes. *J. Mater. Chem. A* **2013**, *1*, 14071–14074. [CrossRef]
24. Fang, W.; Liu, L.; Li, T.; Dang, Z.; Qiao, C.; Xu, J.; Wang, Y. Electrospun N-substituted polyurethane membranes with self-healing ability for self-cleaning and oil/water separation. *Chemistry* **2016**, *22*, 878–883. [CrossRef]
25. Gui, X.C.; Wei, J.Q.; Wang, K.L.; Cao, A.Y.; Zhu, H.W.; Jia, Y.; Shu, Q.K.; Wu, D.H. Carbon nanotube sponges. *Adv. Mater.* **2010**, *22*, 617–621. [CrossRef]
26. Wei, Y.B.; Qi, H.; Gong, X.; Zhao, S.F. Specially wettable membranes for oil-water separation. *Adv. Mater. Inter.* **2018**, 1800576. [CrossRef]
27. Yang, Y.; Tong, Z.; Ngai, T.; Wang, C.Y. Nitrogen-rich and fire-resistant carbon aerogels for the removal of oil contaminants from water. *ACS Appl. Mater. Inter.* **2014**, *6*, 6351–6360. [CrossRef]
28. Song, W.; Zhang, J.J.; Xie, Y.F.; Cong, Q.; Zhao, B. Large-area unmodified superhydrophobic copper substrate can be prepared by an electroless replacement deposition. *J. Colloid Interface Sci.* **2009**, *329*, 208–211. [CrossRef]
29. Chang, F.M.; Cheng, S.L.; Hong, S.J.; Sheng, Y.J.; Tsao, H.K. Superhydrophilicity to superhydrophobicity transition of CuO nanowire films. *Appl. Phys. Lett.* **2010**, *96*, 114101. [CrossRef]
30. Wang, F.J.; Lei, S.; Xue, M.S.; Ou, J.F.; Li, W. In Situ separation and collection of oil from water surface via a novel superoleophilic and superhydrophobic oil containment boom. *Langmuir* **2014**, *30*, 1281–1289. [CrossRef]
31. Ding, Y.B.; Li, Y.; Yang, L.L.; Li, Z.Y.; Xin, W.H.; Liu, X.; Pan, L.; Zhao, J.P. The fabrication of controlled coral-like Cu_2O films and their hydrophobic property. *Appl. Surf. Sci.* **2013**, *266*, 395–399. [CrossRef]
32. Singh, V.; Sheng, Y.J.; Tsao, H.K. Facile fabrication of superhydrophobic copper mesh for oil/water separation and theoretical principle for separation design. *J. Taiwan Inst. Chem. E* **2018**, *87*, 1–8. [CrossRef]
33. Monte, M.; Munuera, G.; Costa, D.; Conesa, J.C.; Martinez, A.A. Near-ambient XPS characterization of interfacial copper species in ceria-supported copper catalysts. *Phys. Chem. Chem. Phys.* **2015**, *17*, 29995–30004. [CrossRef] [PubMed]
34. Li, B.; Luo, X.; Zhu, Y.; Wang, X. Immobilization of Cu(II) in KIT-6 supported Co_3O_4 and catalytic performance for epoxidation of styrene. *App. Surf. Sci.* **2015**, *359*, 609–620. [CrossRef]
35. Embden, J.V.; Tachibana, Y. Synthesis and characterisation of famatinite copper antimony sulfide nanocrystals. *J. Mater. Chem.* **2012**, *22*, 11466–11469. [CrossRef]
36. Lv, Y.Q.; Shi, B.Z.; Qi, Y.; Su, X.F.; Liu, L.; Tian, L.H.; Ding, J. Synthesis and characterization of the maize-leaf like Cu_2O nanosheets array film via an anodic oxidation method. *J. Alloy. Compd.* **2019**, *773*, 706–712. [CrossRef]

37. Cheng, Z.P.; Xu, J.M.; Zhong, H.; Chu, X.Z.; Song, J. Repeatable synthesis of Cu$_2$O nanorods by a simple and novel reduction route. *Mater. Lett.* **2011**, *65*, 1871–1874. [CrossRef]
38. Zhou, C.L.; Cheng, J.; Hou, K.; Zhu, Z.T.; Zheng, Y.F. Preparation of CuWO$_4$@Cu$_2$O film on copper mesh by anodization for oil/water separation and aqueous pollutant degradation. *Chem. Eng. J.* **2017**, *307*, 803–811. [CrossRef]
39. Ai, Z.H.; Zhang, L.Z.; Lee, S.C.; Ho, W.K. Interfacial hydrothermal synthesis of Cu@Cu$_2$O core-shell microspheres with enhanced visible-light-driven photocatalytic activity. *J. Phys. Chem. C* **2009**, *113*, 20896–20902. [CrossRef]
40. Platzman, I.; Brener, R.; Haick, H.; Tannenbaum, R. Oxidation of polycrystalline copper thin films at ambient conditions. *J. Phys. Chem. C* **2008**, *112*, 1101–1108. [CrossRef]
41. Cassie, A.B.D.; Baxter, S. Wettability of porous surfaces. *Trans. Faraday Soc.* **1944**, *40*, 546–551. [CrossRef]
42. Manoudis, P.N.; Karapanagiotis, I. Modification of the wettability of polymer surfaces using nanoparticles. *Prog. Org. Coat.* **2014**, *77*, 331–338. [CrossRef]
43. Lu, Y.; Sathasivam, S.; Song, J.; Chen, F.; Xu, W.; Carmalt, C.J.; Parkin, I.P. Creating superhydrophobic mild steel surfaces for water proofing and oil-water separation. *J. Mater. Chem. A* **2014**, *2*, 11628–11634. [CrossRef]
44. Li, J.; Yan, L.; Li, H.; Li, W.; Zha, F.; Lei, Z. Underwater superoleophobic palygorskite coated meshes for efficient oil/water separation. *J. Mater. Chem. A* **2015**, *3*, 14696–14702. [CrossRef]

 © 2019 by the authors. Licensee MDPI, Basel, Switzerland. This article is an open access article distributed under the terms and conditions of the Creative Commons Attribution (CC BY) license (http://creativecommons.org/licenses/by/4.0/).

Article

Assembly Mechanism and the Morphological Analysis of the Robust Superhydrophobic Surface

Doeun Kim [1,2], Arun Sasidharanpillai [1,2], Ki Hoon Yun [1,3], Younki Lee [2], Dong-Jin Yun [4], Woon Ik Park [1], Jiwon Bang [1] and Seunghyup Lee [1,*]

1 Electronic Convergence Materials Division, Korea Institute of Ceramic Engineering and Technology, Jinju, Gyeongnam 52851, Korea
2 Department of Materials Engineering and Convergence Technology, Gyeongsang National University, Jinju, Gyeongnam 52828, Korea
3 Department of Convergence engineering, Pusan National University, Busan 46241, Korea
4 Analytical Engineering Group, Samsung Advanced Institute of Technology, Suwon 440-600, Korea
* Correspondence: shbelly@kicet.re.kr; Tel.: +82-55-792-2678

Received: 21 June 2019; Accepted: 24 July 2019; Published: 26 July 2019

Abstract: Robust superhydrophobic surfaces are fabricated on different substrates by a scalable spray coating process. The developed superhydrophobic surface consists of thin layers of surface functionalized silica nanoparticle (SiO_2) bound to the substrate by acrylate-polyurethane (PU) binder. The influence of the SiO_2/PU ratio on the superhydrophobicity, and the robustness of the developed surface, is systematically analyzed. The optimized SiO_2/PU ratio for prepared superhydrophobic surfaces is obtained between 0.9 and 1.2. The mechanism which yields superhydrophobicity to the surface is deduced for the first time with the help of scanning electron microscopy and profilometer. The effect of mechanical abrasion on the surface roughness and superhydrophobicity are analyzed by using profilometer and contact angle measurement, respectively. Finally, it is concluded that the binder plays a key role in controlling the surface roughness and superhydrophobicity through the capillary mechanism. Additionally, the reason for the reduction in performance is also discussed with respect to the morphology variation.

Keywords: robust superhydrophobic surface; surface assembly mechanism; surface disintegration mechanism

1. Introduction

Recently, studies regarding biomimetic have been actively carried out as attempts to advance material technology. Super-strong fibers mimicking spider fibers and anti-reflective displays mimicking moth eyes are representative examples. Among them, smart surfaces, which mimic the self-cleaning ability of the lotus, have attracted scientific attention for the past few decades [1–3]. It is generally accepted that superhydrophobic surfaces—contact angle >150° and hysteresis <10°—are able to bounce off water droplets without wetting its surface [4–7]. Efforts to mimic natural superhydrophobicity has lead researchers to fabricate artificial superhydrophobic surfaces which can be applied for a variety of applications, including self-cleaning window and panels, anti-icing, water-resistant fabrics, anti-fouling, drag reduction, corrosion resistance, etc., [8–12]. These properties are used to provide waterproof properties to electronic devices [10], or to provide self-cleaning and anti-fouling functions that remove dirt, viruses, and bacteria from medical apparatuses [13].

The essential requirements for a surface to be superhydrophobic is that it should have low surface energy and complex nano-scale surface morphology as suggested by the Cassie-Baxter model [14–18]. Studies have demonstrated that solid surface with increased surface roughness can enhance superhydrophobicity [6,19]. The rough structure can trap air inside, and pushes off the water droplets which comes in contact with the surface. The underlying mechanism is based on the extreme reduction in contact surface area due to the nanostructure, which helps easy sliding of water droplets [20,21]. Thus, it is necessary to tailor a rough structure for achieving superhydrophobicity.

In order to control surface roughness, several methods have been adopted, including the sol–gel process, chemical etching, layer-by-layer assembly, electrostatic spinning, chemical vapor deposition, and lithographic process [2,17–23] However, these methods have issues, such as vulnerable mechanical strength, process complexity, high investment costs, limited substrate selection, and difficulty in mass production. These issues have been raised, but in particular, poor durability in mechanical strength have been emphasized. Remarkable methods have been proposed for enhancing the mechanical properties of superhydrophobic surfaces [24]. One of these methods is the protection of fragile fine-scale surface topographies against mechanical wear by assembling larger scale sacrificial micro-pillars on the polymer surfaces. When the micro-scale structures are built upon surfaces with nano-level roughness, the durability of compression could be improved up to 120 kPa [25]. However, the production process for such structures is complicated, as the self-cleaning function is affected by the ratio and patterns of the hierarchical structures. Meanwhile, studies using the pyramid structure have indicated that superhydrophobicity is retained after mechanical abrasion. When mechanically worn, the nano-roughness is ensured by sacrificing the tips of pyramids [26]. However, the surface is vulnerable under large and continuous stress. To date, the demand for the robustness of the superhydrophobic surface has not subsided. Recently, a smart approach to increase mechanical durability has been proposed by applying superhydrophobic paint on commercial adhesives to implement properties on various substrates, and to promote robustness [27,28]. This is a simple and flexible method that can be used for mass-production. Even though there are plenty of reports about spray-coating of superhydrophobic surfaces on various substrates [29–31], the study lacks knowledge about the assembly mechanism in the early stage of surface formation. In this work, we synthesized superhydrophobic surfaces by spray coating surface treated SiO_2 agglomerates on dissolved polyurethane (PU) (binder) which are deposited on various substrates. In order to investigate the assembly mechanism, the morphology of the superhydrophobic surfaces was controlled systematically by adjusting the weight ratio of the sprayed particles and the binder. The prepared samples were mechanically abraded by using a rubbing machine. The surface with optimal SiO_2/PU ratio presented robustness that maintained its superhydrophobicity after many cycles of mechanical abrasion. The assembly mechanism responsible for superhydrophobicity in these coatings were studied with the help of scanning electron microscopy and roughness analysis. Furthermore, the degradation in superhydrophobicity (as a result of the collapse of the micro-protuberances) is due to the applied stress is also discussed through roughness analysis. The underlying phenomenon for the formation of the superhydrophobic surface is the creation of two-length scale rough surface on the spray-coated substrate. All of the previous studies analyzed only the effect of particle/polymer ratio on the creation of such surfaces, could not shed light into the mechanism of how these micro/nano roughness are created on the surface of dissolved binder. In this article we report the experimental evidence of binder penetration through the porous SiO_2 agglomerates, and through capillary mechanism, which is responsible for the formation of a robust rough surface. The systematic investigation on the assembly mechanism and the behavior of the sprayed SiO_2 particles presented here will help industrialists to develop mechanically durable superhydrophobic surfaces.

2. Materials and Methods

2.1. Preparation of Hydrophobic Solution

A superhydrophobic solution was prepared through different steps. Initially, 2 g of SiO_2 powder (5–15 nm, Sigma Aldrich, Saint Louis, MO, USA) and 40 mL of absolute ethanol were mixed in a beaker. Since SiO_2 is hydrophilic in nature, surface treatment should be done to make it hydrophobic. It was done by adding 2 mL of Octadecyltrimethoxysilane (OTS, 90%, Sigma Aldrich) into the stirring solution, and the stirring was continued for 2 h. The mixed solution was then poured on to a glass dish and allowed to dry at room temperature for 8 h. The superhydrophobic solution was obtained by dispersing the dried powder in 100 mL of absolute ethanol for 3 h.

2.2. Preparation of Superhydrophobic Coating

Glass slide, aluminum sheet, alumina plate, and artificial fibers were used as the substrate for fabricating superhydrophobic surfaces. For cleaning, all substrates were ultra-sonicated successively for 10 min in ethanol and distilled water, then dried with nitrogen gas. The acrylic polyurethane (PU) resin and curing agent (HS Clear, KCC, Seoul, Korea) prepared in a ratio of 2:1 was stirred manually. The substrate was then spin-coated at 200 RPM with the dissolved PU for 30 s. Immediately, the hydrophobic solution was spray-coated on the surface of the PU coated substrate. In addition, each surface was constructed by systematically adjusting the amount of spray of the superhydrophobic solution. The sample was then cured for 30 min in a hot air oven. The weight of the binder and the sprayed particles was measured after drying completely.

2.3. Characterization

For particle size analysis, the solution was diluted in ethanol and evaluated by measuring contact angle using laser scattering particle size distribution analyzer (LA-950V2, HORIBA, Kyoto, Japan). Superhydrophobicity of the spray-coated substrates were confirmed by contact angle analyzer (Phoenix-10, SEO, Kromtek Sdn Bhd, Shah Alam, Malaysia). The advancing and receding angles of the water droplets on each surface were measured, and the hysteresis was calculated. The surface morphology of each sample was characterized by SEM (SM-300, TOPCON, Tokyo, Japan). Furthermore, the surface roughness measurement of each sample was carried out by Surface Profilometer (Dektak, VEECO, Plainview, NY, USA). For the evaluation of the mechanical durability of superhydrophobic surfaces, the abrasion rubbing machine (KP-M4250, KIPAE, Gyeongju, Korea) was used. The coated substrates under test were fixed on a leveling plate and loaded with a regulated load using a home-made device. We controlled the pressure on coated substrates and measured the changes in contact angle after the abrasion. Ultimately, the durability of the coated surface was identified.

3. Results and Discussion

The fabrication process for hierarchical superhydrophobic surfaces was schematically shown in Figure 1. [32,33]. Particle size analysis of the surface treated and dried SiO_2 agglomerates are shown in Figure 1a. In order to make SiO_2 hydrophobic, the agglomerated powder should be surface functionalized with some suitable agent. Sriramulu et al. has investigated the mechanism of functionalization of silane coupling agent, such as OTS on SiO_2 nanoparticles, for application in superhydrophobic surfaces. They demonstrated that surface functionalization of silica nanoparticle reduces the adsorption of water [34]. The alkylalkoxysilanes like OTS can replace the alkoxy group with –OH in the polar solvent. It requires hydroxylated surface as a substrate for their association. The silanes on the surface of SiO_2 replaced with –OH is adsorbed on the substrate, and then they form Si–O–Si bonds through dehydration and polymerization [35,36]. Through the reaction with OTS, the surface energy of the SiO_2 reduced and hydrophobicity is increased [7]. These surface functionalized SiO_2 agglomerates are sprayed onto the dispersed binder film on the substrate and then cured at 60 °C for 30 min. Applying this technique, the surfaces were fabricated by controlling roughness and hydrophobicity on various substrates.

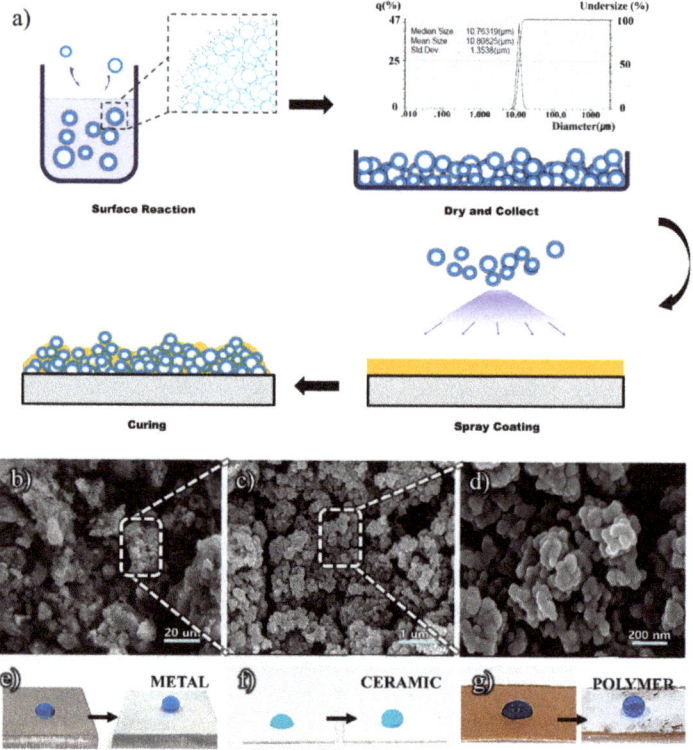

Figure 1. (a) Schematic for the evolution of rough surface using the spray-coating technique, (**b**–**d**) Surface morphology of the superhydrophobic surface structure at different magnifications, (**e**–**g**) Digital images of water droplets on the various superhydrophobic surfaces.

Figure 1b–d shows the surface morphologies of the superhydrophobic surface at different magnifications. It is clear from the figure that fine nanoparticles are clustered to form a surface with higher roughness, which is considered as one of the essential requirements for superhydrophobicity. Additionally, the observed porosity inside the structure can trap air and thereby reduce the contact area of water drops with the surface, which is critical for enhancing superhydrophobicity. In short, superhydrophobicity could be achieved by using hierarchical arrangements of micro/nano-sized structure. Ironically, poor dispersion of SiO_2 nanoparticle is preferably effective in making such a rough hierarchical surface. The proposed technique for making superhydrophobic surface has greater advantages over conventional methods—in this case, there is a wider possibility of application of any type of solid substrate. It is believed that superhydrophobicity is only affected by the assembly of the binder and surface modified particles. For example, on a variety of substrates made of materials, such as metal, ceramic, and polymers, superhydrophobicity was successfully obtained, and is shown in Figure 1e–g. There is no shape change observed for ceramic substrates, due to its inherent rigidity and stiffness. This method is applicable to any substrate to which the binder has good adhesion. In our work, we used acrylate-containing PU as a binder, which has many advantages, including good mechanical properties, low-temperature curability, solvent resistance, etc., [37,38].

We systematically investigated the influence of filler and binder on the roughness of the coated substrates by controlling the powder to binder weight ratio. Figure 2a–d shows the SEM image and roughness profile of the spray-coated substrates with different SiO_2/PU ratio (denoted as R_0.03, R_0.15, R_0.3, R_0.9). The weight of the PU coating was measured in an indirect way—by subtracting the weight of the substrate before coating from that of the substrate after coating. The same procedure was adopted for measuring the weight of the SiO$_2$ nanopowder. It should be noted that all the weights were taken on dried samples. After analyzing the SEM images (Figure 2a), it is clear that the low SiO$_2$/PU weight ratio (R_0.03) is not enough to achieve a high roughness for getting superhydrophobicity. The measured roughness for R_0.03 is 0.116 µm, the corresponding contact angle is 74.7°, which is well below the par level to achieve superhydrophobicity. Also, the particles are seen almost immersed in the binder matrix, as a result of the action of binder to minimize its surface energy [19,39]. Moving on to Figure 2b, the increased weight ratio (R_0.15) has improved the surface roughness to a much higher value of 3.921 µm, still, the contact angle is very low (90.8°) to provide superhydrophobicity. Increasing weight ratio to higher values, for example, to 0.3 and 0.9, could enhance the surface roughness to 13.165 µm for R_0.3 and 15.517 µm for R_0.9, respectively. Corresponding contact angles are 148.2° and 158°, which clearly indicates that the surface is superhydrophobic. In Figure 2d, a large scale-roughness with 15.517 µm was detected, but the change of the contact angle and R_a was reduced. The stage of R_0.3 and R_0.9 show the surface completely covered by particles. Thus, the surface roughness increases by the number of the sprayed particle, but tends to be no longer increased above a certain level.

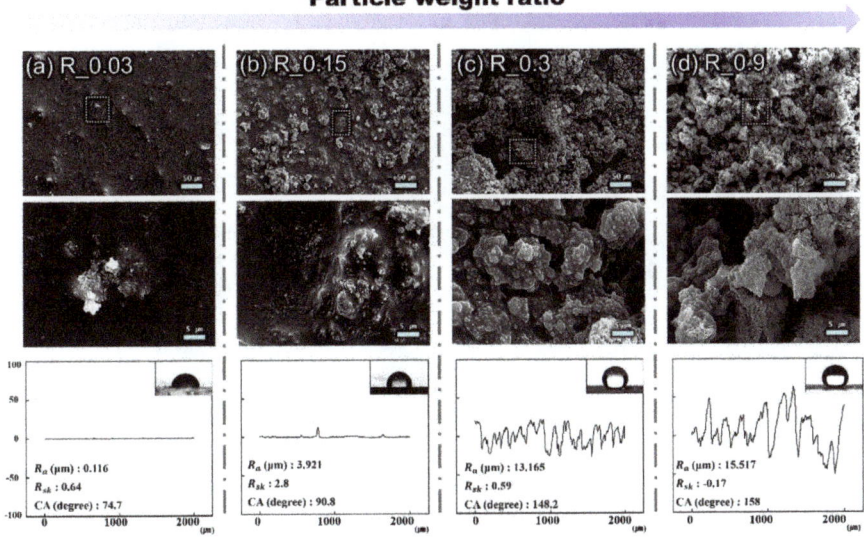

Figure 2. SEM images, roughness profiles and contact angle images of the surface with different weight ratios of the particles to the binder of (**a**) R_0.03, (**b**) R_0.15, (**c**) R_0.3, and (**d**) R_0.9.

The schematics for the SiO$_2$ nanoparticle assembly mechanism on PU are shown in Figure 3. When agglomerated SiO$_2$ nanoparticles sprayed on to the surface of the binder, the binder ascends and holds the nanoparticles. As explained earlier, this occurs as a result of the action of binder to minimize surface energy [40,41]. Meanwhile, to minimize the surface energy, it is confirmed that the PU covers the particles (SEM image of Figure 2a). Creation of multimodal texture, is essential for fabrication of superhydrophobic surface with particles having lower hydrophobicity, can be done either by aggregation of particles directly on the substrate or by deposition of dispersed aggregates. Conventionally, these dispersions are directly coated on the substrate. Thus, the peculiarities of the formed structures are the result of the interplay between various forces, such as van der Waals interactions, depletion interaction, ion-electrostatic repulsion and ultimately the image-charge forces and the structural forces, particularly its polarity and magnitude determine the properties of the final system. Since superhydrophobicity is a surface feature, which depends only on the nature of a few monolayers on the surface, we have the freedom to choose any material which can strongly bond the surface layer with the substrate [32]. In our experiment, we used strongly adhering polyurethane coating on the substrates and sprayed the highly agglomerated and surface functionalized SiO_2 nanoparticles (Agglomerated average particle size = 10 µm) dispersed in ethanol (Figure 1). There is a sharp rise in the binder level above its mean surface, as shown in the cross-sectional SEM image in Figure 3, which let us attribute capillary action of the pre-coated binder as the responsible mechanism for creating such a rough surface with plenty of micro-protuberances. Because of the fine space between the sprayed particles, the binder can effectively infiltrate and surround. Due to the surface tension of the penetrated binder, it was dragged up to the top of the SiO_2 agglomerates. This continuous process would finally yield a rough surface with plenty of hills and valleys, which can trap air inside, and, thus, forms a perfect superhydrophobic surface.

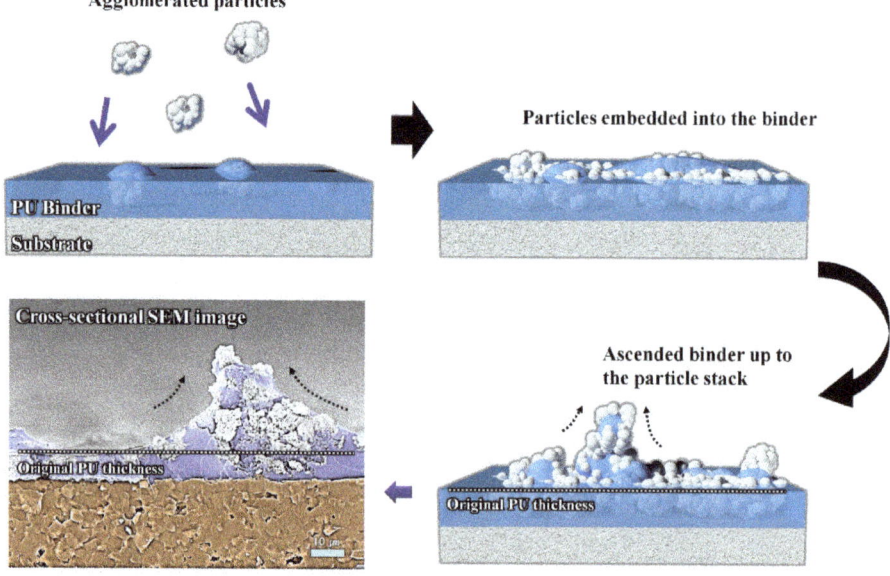

Figure 3. Schematic image of the surface structure assembly mechanism and the fractured SEM image of the sprayed superhydrophobic surface (False color).

One of the obstacles preventing superhydrophobic surfaces from its application in many fields is its poor stability against abrasive wear [42]. It is known that during the friction cycle, the micro-protuberances are easily broken, the surface becomes smoother, and eventually superhydrophobicity will be lost. Polyurethane (PU) based elastomers have a successful history of application on improving mechanical properties of superhydrophobic surfaces, since they can retain the structure even after abrasion. Changhong et al. have fabricated superhydrophobic surfaces on porous Al template using PU elastomers, they could keep superhydrophobicity even after 10,000 cycles of rubbing at 18 cm·s^{-1} with a load of 2945.7 Pa [43]. As evident from Figure 3, the SiO_2 nanoparticles are completely surrounded by PU binder. The presence of binder in the composite structure can resist or partially absorb mechanical stresses applied to it. The durability of the superhydrophobic surface was evaluated through abrasion testing. Milionois et al. and Bayer et al. have extensively reviewed the available superhydrophobic materials and commented on their mechanical integrity. There are various techniques to analyze mechanical durability of a superhydrophobic surface, such as adhesion tape peeling, sand abrasion test and tangential abrasion test (Table 1) [11,44,45]. In this work, we performed a tangential mechanical abrasion test to evaluate the durability of coating over several cycles of abrasion.

Table 1. Comparison of abrasion wears resistance with the previous reports on the superhydrophobic surface based on spray coating of composited materials.

Superhydrophobic Material/Method	Total Abrasion Cycles	Load	Reference
MoS_2/PU (spray coating)	Over 100 m rubbing distance	500 g	Tang et al. [46]
SiO_2/Starch (spray coating)	17	1.1–2.1 kPa	Milionis et al. [47]
SiO_2/Siloxane (spray coating)	10	<10 kPa	Chen et al. [48]
SiO_2/EAC (spray coating)	10	40 kPa	Tenjimbayashi and Shiratori [49]
Grephene/PU (spray coating)	30	15 kPa	Naderizadeh et al. [45]
SiO_2/PU (spray coating)	100	3.138 kPa	Our study

For the evaluation, the advancing and receding contact angle was measured before and after the abrasion. The rubbing test was conducted, as shown in the inset of Figure 4, applying 3.138 kPa load on it. In Figure 4, the effect of abrasion on superhydrophobicity was plotted for surfaces with varying powder/binder ($R\_$) ratio. It is primarily understood from the figure that useful results are obtained only for surfaces with $R\_$ from 0.9 to 1.2. In the proposed range, the advancing contact angle is 158°–159°, and the hysteresis is 4°–5° before rubbing. Surprisingly, even after rubbing, the advancing contact angle is between 153° and 154°, and hysteresis between 7° and 9°. One of the criteria for superhydrophobic surfaces–hysteresis less than 10° is well maintained from 0.9 to 1.2. It is only possible in a range of ensuring durability by compositing the surface structure mentioned above. For $R\_$ values less than 0.9 and greater than 1.2, there is a big difference between advancing and receding angle before and after abrasion. In surfaces with lower $R\_$ ratios, as clear from Figure 2, the roughness is too low. For $R\_0.2$, there was a sudden fall in advancing angle after abrasion from 147° to around 125°, which cannot retain superhydrophobicity. After abrasion, the advancing angle of $R\_0.4$ reduced from 155.6° to 147.2° with a hysteresis of 4.6°. Even a very small change in protuberance height can offer a very high fall in contact angle. For higher $R\_$ values, $R\_$ greater than 1.2 all substrates showed superhydrophobicity before rubbing (advancing angles $\geq$ 158°, hysteresis $\leq$ 6°). However, it cannot maintain non-wetting after abrasion, and, consequently, the measured advancing angles of $R\_1.3$ and $R\_2.1$ reduced to 141° and 132°, respectively. Loss of superhydrophobicity is expected due to the poor attachment of SiO_2 nanoparticles with the PU binder, since the spray-coated thickness is above the penetration level of binder. Hence, any type of mechanical wear can wash away the surface roughness and make it flatter. A second observation is that irrespective of the $R\_$ ratio, abrasion has reduced the contact angle of all substrates. In the given fabrication mechanism, even if the structure is broken, due to the destruction of the binder by abrasion, the SiO_2 particles can be exposed on the surface, and the roughness and hydrophobicity can be maintained. Thus, the durability of the surface against mechanical abrasion can be improved to higher levels than the conventional nanostructured superhydrophobic surface. A comparison of the abrasion test with other available literature data are presented in Table 1: The obtained results are promising and comparable with the literature data.

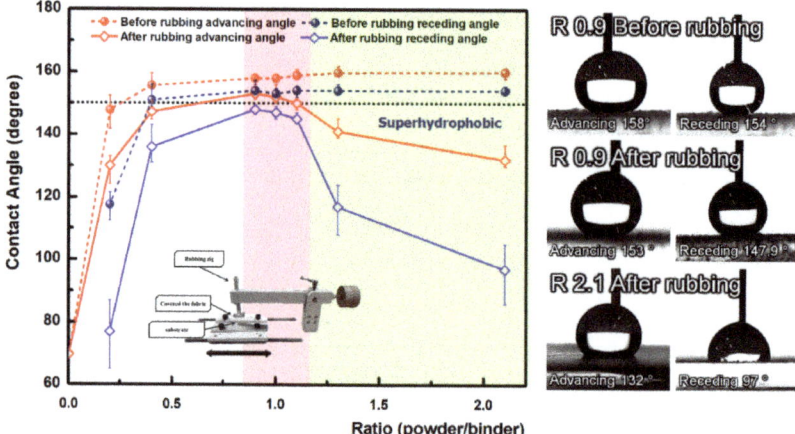

Figure 4. Advancing and receding contact angles with different particle weight ratio surfaces, before and after mechanical abrasion test.

For an in-depth investigation of mechanical abrasion on change in surface structure and performance degradation, we analyzed the degraded surface by SEM and profilometer. Generated roughness and degradation in superhydrophobicity can be evaluated by measuring average surface roughness, R_a and skewness, R_{sk}. The R_a, average roughness, is the arithmetic average of the absolute values of the profile heights over the evaluation length. Skewness is a measure of the symmetry of height distribution and provides details about the number of hills and valleys on the surface. R_a and R_{sk} can be calculated by the following equations [50],

$$R_a = \frac{1}{N}\sum_{i=1}^{N}|Z_i - \overline{Z}| \tag{1}$$

$$R_{sk} = \frac{1}{R_q^3}\frac{1}{N}\sum_{i=1}^{N}(Z_i - \overline{Z})^3 \tag{2}$$

where N equal to 18,000, is the total number of measured points for a scan length of 2 mm, R_q is root-mean-square (RMS) roughness which is calculated by the profilometer software from the measured height data, $\overline{Z}$ is the mean-height distance, and surface height data Z_i obtained from scan area. SEM images of Figures 2a–d and 5c,d indicate that with an increase in the amount of sprayed articles, the surface roughness also increases. Due to the increased roughness, the number of air pockets also increases, and superhydrophobicity is achieved. However, surface roughness (R_a) itself is not a suitable parameter to explain the superhydrophobicity—we should depend on other factors, such as skewness (R_{sk}), for a complete interpretation. Negative skewness indicates a greater percentage of the profile is above the mean line, and a positive value indicates a greater percentage is below the mean line. When the absolute value of R_{sk} approaches zero, the shape of the surface tends to be regular and symmetric. For example, if R_{sk} is a high positive value, it would tend to have a large number of hills, and few valleys on the surface [50,51]. Since the basic strategy behind constructing superhydrophobic surfaces are creating microscale rough structures with long range order and symmetry, skewness can provide more information on this phenomenon.

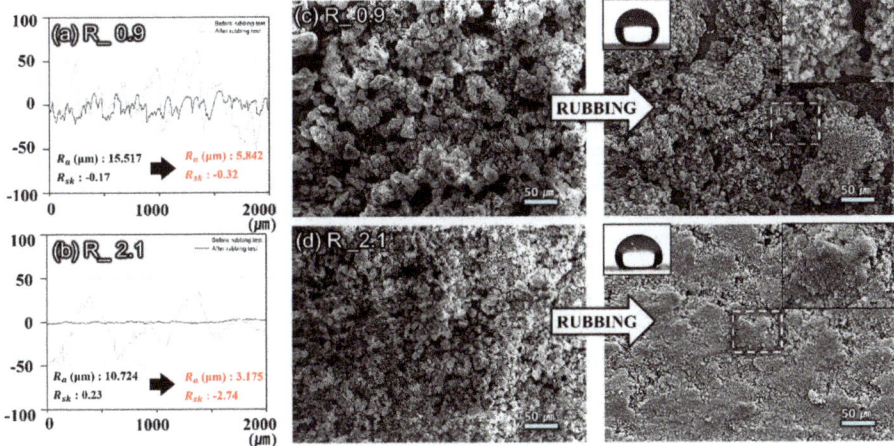

Figure 5. Roughness profiles before and after rubbing the surface with (**a**) R_0.9 (**b**) R_2.1 and SEM images before and after rubbing and contact angle images after rubbing the surface with (**c**) R_0.9 (**d**) R_2.1.

Figure 5a,b shows the roughness data of the surface of R_0.9 and R_2.1 before and after abrasion test. For R_0.9, even though the roughness has undergone a threefold reduction from 15.517 to 5.842 μm, due to the abrasion, the comparatively similar skewness tells us that the profile is still symmetric and regular. Since the difference in R_{sk} before and after abrasion is negligible, the superhydrophobicity could be maintained [52]. These observations are in well agreement with the surface morphology of R_0.9 (Figure 5c). By contrast, R_2.1 shows a large change in R_a and R_{sk} after rubbing. A remarkable reduction in R_a from 10.7 to 3.17 μm is observed, whereas R_{sk} reached a very low value of −2.74 after prolonged rubbing. The reason for the decreased contact angle can be attributed to the value of skewness. As described earlier, a negative R_{sk} means that the roughness parameter is tilted up. Since R_{sk} shows high negative value, it could be predicted that the surface had few numbers of deep valleys and relatively flat top surface. It is consistent with the SEM images of Figure 5d. In fact, in the analysis by SEM observation, there are no more protuberances observed on the surface of R_2.1 after rubbing (Figure 5d), and a flattened surface is obtained. The flat surface with few deep valleys is not accounted for by the collapse of surface structures as reported previously. This flatness is the result of the removal of particles from the top of the hills and its deposition in the deep valleys. Since R_2.1 contains a quantity of SiO_2 powder greater than its accommodation limit, there can be few layers of particles which are ahead of the penetration limit of PU, are loosely attached. On rubbing, these particles are easily removed from the top and deposited inside the deeper valleys. This process reduces the number of air-pockets, and, hence, suppresses superhydrophobicity, as it depends on the number of air cavities on the top surface. This observation is matching with the SEM image of Figure 5d.

To summarize, the binder, as mentioned in the description of the mechanism, penetrated through the particles, mixed, and combined to form a composite material, which could resist stress and support the particles. With this mechanism, we fabricated a robust superhydrophobic surface based on the penetration of polymer and organic solvent.

4. Conclusions

This paper discusses the fabrication of superhydrophobic surface by spray coating functionalized SiO_2 nanoparticle on PU binder on a different substrate. The robustness of the coating against mechanical abrasions was tested, and its effect on superhydrophobicity is analyzed by contact angle measurement with the support of surface profilometer and SEM analysis. We systematically investigated the influence of the weight fractions of surface functionalized SiO_2 nanoparticle and PU binder on the superhydrophobic performance of the surface. The capillary action of the dissolved binder through the agglomerated particle is observed and is responsible for holding the particle tightly and keep it protected against mechanical abrasion. This particular behavior of the binder is first identified in our work. The samples with specific ratio demonstrated a contact angle greater than 150° after 100 abrasion cycles, and the optimum ratio for maintaining the surface structure against mechanical abrasion is obtained between 0.9 and 1.2. The relatively high abrasion resistance is believed to be due to the influence of the mechanically strong PU binder employed here. In addition, the principle of performance degradation and the collapse of the surface structure were studied through the roughness analysis. Over the optimized particle ratio, interesting surface changes, due to collapse is observed after rubbing. In particular, a flat surface structure was generated by the unsupported and excessive particle lumps. As the structure changed, the superhydrophobic performance was lowered as well. The design and manufacture with the proposed particle-to-binder ratio could increase the mechanical durability of the superhydrophobic surface.

Author Contributions: Conceptualization, D.K. and S.L.; Methodology, D.K. and A.S.; Resources, D.K., A.S. and K.H.Y.; Formal analysis, D.-J.Y.; Investigation, D.K., W.I.P. and J.B.; Data curation, D.-J.Y.; Visualization, K.H.Y. and J.B.; Writing—Original Draft Preparation, D.K.; Writing—Review and Editing, A.S. and Y.L.; Supervision, S.L.; Project Administration, S.L.; Funding Acquisition, S.L.

Funding: This work was supported by the National Research Foundation of Korea (NRF, NRF-2017R1D1A1B03035258) grant funded by the Korea government (MSIT) and the Technology Development Program (S2587969) funded by the Ministry of SMEs and Startups (MSS, Korea).

Conflicts of Interest: The authors declare no conflict of interest.

References

1. Lee, B.J.; Zhang, Z.; Baek, S.; Kim, S.; Kim, D.; Yong, K. Bio-inspired dewetted surfaces based on SiC/Si interlocked structures for enhanced-underwater stability and regenerative-drag reduction capability. *Sci. Rep.* **2016**, *6*, 24653. [CrossRef] [PubMed]
2. Yang, Y.; Li, X.; Zheng, X.; Chen, Z.; Zhou, Q.; Chen, Y. 3D-printed biomimetic super-hydrophobic structure for microdroplet manipulation and oil/water separation. *Adv. Mater.* **2018**, *30*, 1704912. [CrossRef] [PubMed]
3. Peng, C.; Chen, Z.; Tiwari, M.K. All-organic superhydrophobic coatings with mechanochemical robustness and liquid impalement resistance. *Nat. Mater.* **2018**, *17*, 355–360. [CrossRef] [PubMed]
4. Jankauskaitė, V.; Narmontas, P.; Lazauskas, A. Control of polydimethylsiloxane surface hydrophobicity by plasma polymerized hexamethyldisilazane deposition. *Coatings* **2019**, *9*, 36.
5. Sakai, M.; Song, J.-H.; Yoshida, N.; Suzuki, S.; Kameshima, Y.; Nakajima, A. Relationship between sliding acceleration of water droplets and dynamic contact angles on hydrophobic surfaces. *Surf. Sci.* **2006**, *600*, L204–L208. [CrossRef]
6. Han, M.H.; Park, Y.H.; Hyun, J.W.; Ahn, Y.H. Facile method for fabricating superhydrophobic surface on magnesium. *Bull. Korean Chem. Soc.* **2010**, *31*, 1067–1069. [CrossRef]
7. Dai, X.; Sun, N.; Nielsen, S.O.; Stogin, B.B.; Wang, J.; Yang, S.; Wong, T.-S. Hydrophilic directional slippery rough surfaces for water harvesting. *Sci. Adv.* **2018**, *4*, eaaq0919. [CrossRef]
8. Karmouch, R.; Ross, G.G. Superhydrophobic wind turbine blade surfaces obtained by a simple deposition of silica nanoparticles embedded in epoxy. *Appl. Surf. Sci.* **2010**, *257*, 665–669. [CrossRef]
9. Feng, L.; Che, Y.; Liu, Y.; Qiang, X.; Wang, Y. Fabrication of superhydrophobic aluminium alloy surface with excellent corrosion resistance by a facile and environment-friendly method. *Appl. Surf. Sci.* **2013**, *283*, 367–374. [CrossRef]
10. Lee, S.; Kim, W.; Yong, K. Overcoming the water vulnerability of electronic devices: A highly water-resistant ZnO nanodevice with multifunctionality. *Adv. Mater.* **2011**, *23*, 4398–4402. [CrossRef]
11. Bayer, S.I. On the durability and wear resistance of transparent superhydrophobic coatings. *Coatings* **2017**, *7*, 12. [CrossRef]
12. Zeng, Y.; Qin, Z.; Hua, Q.; Min, Y.; Xu, Q. Sheet-like superhydrophobic surfaces fabricated on copper as a barrier to corrosion in a simulated marine system. *Surf. Coat. Technol.* **2019**, *362*, 62–71. [CrossRef]

13. Jokinen, V.; Kankuri, E.; Hoshian, S.; Franssila, S.; Ras, R.H.A. Superhydrophobic blood-repellent surfaces. *Adv. Mater.* **2018**, *30*, 1705104. [CrossRef] [PubMed]
14. Pan, Q.; Wang, M.; Wang, H. Separating small amount of water and hydrophobic solvents by novel superhydrophobic copper meshes. *Appl. Surf. Sci.* **2008**, *254*, 6002–6006. [CrossRef]
15. Teisala, H.; Tuominen, M.; Aromaa, M.; Mäkelä, J.M.; Stepien, M.; Saarinen, J.J.; Toivakka, M.; Kuusipalo, J. Development of superhydrophobic coating on paperboard surface using the liquid flame spray. *Surf. Coat. Technol.* **2010**, *205*, 436–445. [CrossRef]
16. Lai, Y.; Gao, X.; Zhuang, H.; Huang, J.; Lin, C.; Jiang, L. Designing superhydrophobic porous nanostructures with tunable water adhesion. *Adv. Mater.* **2009**, *21*, 3799–3803. [CrossRef]
17. Lim, H.S.; Han, J.T.; Kwak, D.; Jin, M.; Cho, K. Photoreversibly switchable superhydrophobic surface with erasable and rewritable pattern. *J. Am. Chem. Soc.* **2006**, *128*, 14458–14459. [CrossRef] [PubMed]
18. Saleema, N.; Sarkar, D.K.; Gallant, D.; Paynter, R.W.; Chen, X.-G. Chemical nature of superhydrophobic aluminum alloy surfaces produced via a one-step process using fluoroalkyl-silane in a base medium. *ACS Appl. Mater. Interfaces* **2011**, *3*, 4775–4781. [CrossRef] [PubMed]
19. Han, J.T.; Xu, X.; Cho, K. Diverse access to artificial superhydrophobic surfaces using block copolymers. *Langmuir* **2005**, *21*, 6662–6665. [CrossRef]
20. Xue, C.H.; Jia, S.T.; Chen, H.Z.; Wang, M. Superhydrophobic cotton fabrics prepared by sol-gel coating of TiO_2 and surface hydrophobization. *Sci. Technol. Adv. Mater.* **2008**, *9*, 35001. [CrossRef]
21. Senesi, G.S.; D'Aloia, E.; Gristina, R.; Favia, P.; d'Agostino, R. Surface characterization of plasma deposited nano-structured fluorocarbon coatings for promoting in vitro cell growth. *Surf. Sci.* **2007**, *601*, 1019–1025. [CrossRef]
22. Pan, G.; Xiao, X.; Yu, N.; Ye, Z. Fabrication of superhydrophobic coatings on cotton fabric using ultrasound-assisted in-situ growth method. *Prog. Org. Coat.* **2018**, *125*, 463–471. [CrossRef]
23. Rivero, J.P.; Iribarren, A.; Larumbe, S.; Palacio, F.J.; Rodríguez, R. A comparative study of multifunctional coatings based on electrospun fibers with incorporated ZnO nanoparticles. *Coatings* **2019**, *9*, 367. [CrossRef]
24. Verho, T.; Bower, C.; Andrew, P.; Franssila, S.; Ikkala, O.; Ras, R.H.A. Mechanically durable superhydrophobic surfaces. *Adv. Mater.* **2010**, *23*, 673–678. [CrossRef] [PubMed]
25. Huovinen, E.; Takkunen, L.; Korpela, T.; Suvanto, M.; Pakkanen, T.T.; Pakkanen, T.A. Mechanically robust superhydrophobic polymer surfaces based on protective micropillars. *Langmuir* **2014**, *30*, 1435–1443. [CrossRef] [PubMed]
26. Xiu, Y.; Liu, Y.; Hess, D.W.; Wong, C.P. Mechanically robust superhydrophobicity on hierarchically structured Si surfaces. *Nanotechnology* **2010**, *21*, 155705. [CrossRef] [PubMed]
27. Lu, Y.; Sathasivam, S.; Song, J.; Crick, C.R.; Carmalt, C.J.; Parkin, I.P. Robust self-cleaning surfaces that function when exposed to either air or oil. *Science* **2015**, *347*, 1132–1135. [CrossRef] [PubMed]
28. Lazauskas, A.; Grigaliūnas, V.; Jucius, D. Recovery behavior of microstructured thiol-ene shape-memory film. *Coatings* **2019**, *9*, 267. [CrossRef]
29. Aslanidou, D.; Karapanagiotis, I.; Panayiotou, C. Superhydrophobic, superoleophobic coatings for the protection of silk textiles. *Prog. Org. Coat.* **2016**, *97*, 44–52. [CrossRef]
30. Karapanagiotis, I.; Manoudis, P.N.; Savva, A.; Panayiotou, C. Superhydrophobic polymer-particle composite films produced using various particle sizes. *Surf. Interface Anal.* **2012**, *44*, 870–875. [CrossRef]
31. Manoudis, P.N.; Karapanagiotis, I.; Tsakalof, A.; Zuburtikudis, I.; Panayiotou, C. Superhydrophobic composite films produced on various substrates. *Langmuir* **2008**, *24*, 11225–11232. [CrossRef] [PubMed]
32. Boinovich, L.; Emelyanenko, A. Principles of design of superhydrophobic coatings by deposition from dispersions. *Langmuir* **2009**, *25*, 2907–2912. [CrossRef] [PubMed]
33. Rahman, I.A.; Padavettan, V. Synthesis of silica nanoparticles by sol-gel: Size-dependent properties, surface modification, and applications in silica-polymer nanocomposites—A review. *J. Nanomater.* **2012**, *2012*, 8. [CrossRef]
34. Sriramulu, D.; Reed, E.L.; Annamalai, M.; Venkatesan, T.V.; Valiyaveettil, S. Synthesis and characterization of superhydrophobic, self-cleaning nir-reflective silica nanoparticles. *Sci. Rep.* **2016**, *6*, 35993. [CrossRef] [PubMed]
35. Ulman, A. Formation and structure of self-assembled monolayers. *Chem. Rev.* **1996**, *96*, 1533–1554. [CrossRef]
36. Sagiv, J. Organized monolayers by adsorption. 1. Formation and structure of oleophobic mixed monolayers on solid surfaces. *J. Am. Chem. Soc.* **1980**, *102*, 92–98. [CrossRef]

37. Rashvand, M.; Ranjbar, Z.; Rastegar, S. Nano zinc oxide as a UV-stabilizer for aromatic polyurethane coatings. *Prog. Org. Coat.* **2011**, *71*, 362–368. [CrossRef]
38. Rosu, D.; Rosu, L.; Cascaval, C.N. IR-change and yellowing of polyurethane as a result of UV irradiation. *Polym. Degrad. Stab.* **2009**, *94*, 591–596. [CrossRef]
39. Gaines, G.L., Jr. Surface and interfacial tension of polymer liquids—A review. *Polym. Eng. Sci.* **1972**, *12*, 1–11. [CrossRef]
40. Dee, G.T.; Sauer, B.B. The surface tension of polymer liquids. *Adv. Phys.* **1998**, *47*, 161–205. [CrossRef]
41. Llaneza, V.; Belzunce, F.J. Study of the effects produced by shot peening on the surface of quenched and tempered steels: Roughness, residual stresses and work hardening. *Appl. Surf. Sci.* **2015**, *356*, 475–485. [CrossRef]
42. Roach, P.; Shirtcliffe, N.J.; Newton, M.I. Progess in superhydrophobic surface development. *Soft Matter* **2008**, *4*, 224–240. [CrossRef]
43. Su, C.; Xu, Y.; Gong, F.; Wang, F.; Li, C. The abrasion resistance of a superhydrophobic surface comprised of polyurethane elastomer. *Soft Matter* **2010**, *6*, 6068–6071. [CrossRef]
44. Milionis, A.; Loth, E.; Bayer, I.S. Recent advances in the mechanical durability of superhydrophobic materials. *Adv. Coll. Interface Sci.* **2016**, *229*, 57–79. [CrossRef] [PubMed]
45. Naderizadeh, S.; Athanassiou, A.; Bayer, I.S. Interfacing superhydrophobic silica nanoparticle films with graphene and thermoplastic polyurethane for wear/abrasion resistance. *J. Coll. Interface Sci.* **2018**, *519*, 285–295. [CrossRef] [PubMed]
46. Tang, Y.; Yang, J.; Yin, L.; Chen, B.; Tang, H.; Liu, C.; Li, C. Fabrication of superhydrophobic polyurethane/MoS$_2$ nanocomposite coatings with wear-resistance. *Coll. Surf. A Physicochem. Eng. Asp.* **2014**, *459*, 261–266. [CrossRef]
47. Milionis, A.; Ruffilli, R.; Bayer, I.S. Superhydrophobic nanocomposites from biodegradable thermoplastic starch composites (Mater-Bi®), hydrophobic nano-silica and lycopodium spores. *RSC Adv.* **2014**, *4*, 34395–34404. [CrossRef]
48. Chen, K.; Zhou, S.; Wu, L. Facile fabrication of self-repairing superhydrophobic coatings. *Chem. Commun.* **2014**, *50*, 11891–11894. [CrossRef]
49. Tenjimbayashi, M.; Shiratori, S. Highly durable superhydrophobic coatings with gradient density by movable spray method. *J. Appl. Phys.* **2014**, *116*, 114310. [CrossRef]
50. Zhang, H.-S.; Endrino, J.L.; Anders, A. Comparative surface and nano-tribological characteristics of nanocomposite diamond-like carbon thin films doped by silver. *Appl. Surf. Sci.* **2008**, *255*, 2551–2556. [CrossRef]
51. Horváth, R.; Czifra, Á.; Drégelyi-Kiss, Á. Effect of conventional and non-conventional tool geometries to skewness and kurtosis of surface roughness in case of fine turning of aluminium alloys with diamond tools. *Int. J. Adv. Manuf. Technol.* **2015**, *78*, 297–304. [CrossRef]
52. Patel, K.; Doyle, C.S.; Yonekura, D.; James, B.J. Effect of surface roughness parameters on thermally sprayed PEEK coatings. *Surf. Coat. Technol.* **2010**, *204*, 3567–3572. [CrossRef]

 © 2019 by the authors. Licensee MDPI, Basel, Switzerland. This article is an open access article distributed under the terms and conditions of the Creative Commons Attribution (CC BY) license (http://creativecommons.org/licenses/by/4.0/).

MDPI
St. Alban-Anlage 66
4052 Basel
Switzerland
Tel. +41 61 683 77 34
Fax +41 61 302 89 18
www.mdpi.com

Coatings Editorial Office
E-mail: coatings@mdpi.com
www.mdpi.com/journal/coatings

www.ingramcontent.com/pod-product-compliance
Lightning Source LLC
LaVergne TN
LVHW070542100526
838202LV00012B/351

9 783039 435418